Metal Nanoparticles-Polymer Hybrid Materials

Metal Nanoparticles-Polymer Hybrid Materials

Editor

Iole Venditti

MDPI • Basel • Beijing • Wuhan • Barcelona • Belgrade • Manchester • Tokyo • Cluj • Tianjin

Editor
Iole Venditti
Roma Tre University
Italy

Editorial Office
MDPI
St. Alban-Anlage 66
4052 Basel, Switzerland

This is a reprint of articles from the Special Issue published online in the open access journal *Polymers* (ISSN 2073-4360) (available at: https://www.mdpi.com/journal/polymers/special_issues/Metal_Nano_Polymers).

For citation purposes, cite each article independently as indicated on the article page online and as indicated below:

LastName, A.A.; LastName, B.B.; LastName, C.C. Article Title. *Journal Name* **Year**, *Volume Number*, Page Range.

ISBN 978-3-0365-5203-3 (Hbk)
ISBN 978-3-0365-5204-0 (PDF)

© 2022 by the authors. Articles in this book are Open Access and distributed under the Creative Commons Attribution (CC BY) license, which allows users to download, copy and build upon published articles, as long as the author and publisher are properly credited, which ensures maximum dissemination and a wider impact of our publications.

The book as a whole is distributed by MDPI under the terms and conditions of the Creative Commons license CC BY-NC-ND.

Contents

About the Editor vii

Preface to "Metal Nanoparticles-Polymer Hybrid Materials" ix

Iole Venditti
Metal Nanoparticles–Polymers Hybrid Materials I
Reprinted from: *Polymers* **2022**, *14*, 3117, doi:10.3390/polym14153117 1

Andrea Fiorati, Arianna Bellingeri, Carlo Punta, Ilaria Corsi and Iole Venditti
Silver Nanoparticles for Water Pollution Monitoring and Treatments: Ecosafety Challenge and Cellulose-Based Hybrids Solution
Reprinted from: *Polymers* **2020**, *12*, 1635, doi:10.3390/polym12081635 5

Irena Jacukowicz-Sobala, Ewa Stanisławska, Agnieszka Baszczuk, Marek Jasiorski and Elżbieta Kociołek-Balawejder
Size-Controlled Transformation of Cu_2O into Zero Valent Copper within the Matrix of Anion Exchangers via Green Chemical Reduction
Reprinted from: *Polymers* **2020**, *12*, 2629, doi:10.3390/polym12112629 29

Gina M. DiSalvo, Abby R. Robinson, Mohamed S. Aly, Eric R. Hoglund, Sean M. O'Malley and Julianne C. Griepenburg
Polymersome Poration and Rupture Mediated by Plasmonic Nanoparticles in Response to Single-Pulse Irradiation
Reprinted from: *Polymers* **2020**, *12*, 2381, doi:10.3390/polym12102381 51

Woong Gi Lee, Younghyun Cho and Sang Wook Kang
Effect of Ionic Radius in Metal Nitrate on Pore Generation of Cellulose Acetate in Polymer Nanocomposite
Reprinted from: *Polymers* **2020**, *12*, 981, doi:10.3390/polym12040981 65

Julia Radwan-Pragłowska, Łukasz Janus, Marek Piątkowski, Dariusz Bogdał and Dalibor Matysek
3D Hierarchical, Nanostructured Chitosan/PLA/HA Scaffolds Doped with TiO_2/Au/Pt NPs with Tunable Properties for Guided Bone Tissue Engineering
Reprinted from: *Polymers* **2020**, *12*, 792, doi:10.3390/polym12040792 75

Meimei Wu, Chao Zhang, Yihan Ji, Yuan Tian, Haonan Wei, Chonghui Li, Zhen Li, Tiying Zhu, Qianqian Sun, Baoyuan Man and Mei Liu
3D Ultrasensitive Polymers-Plasmonic Hybrid Flexible Platform for In-Situ Detection
Reprinted from: *Polymers* **2020**, *12*, 392, doi:10.3390/polym12020392 91

Jorge A Roacho-Pérez, Fernando G Ruiz-Hernandez, Christian Chapa-Gonzalez, Herminia G Martínez-Rodríguez, Israel A Flores-Urquizo, Florencia E Pedroza-Montoya, Elsa N Garza-Treviño, Minerva Bautista-Villareal, Perla E García-Casillas and Celia N Sánchez-Domínguez
Magnetite Nanoparticles Coated with PEG 3350-Tween 80: In Vitro Characterization Using Primary Cell Cultures
Reprinted from: *Polymers* **2020**, *12*, 300, doi:10.3390/polym12020300 105

Peiyuan Shao, Peng Xu, Lei Zhang, Yun Xue, Xihui Zhao, Zichao Li and Qun Li
Non-Chloride in Situ Preparation of Nano-Cuprous Oxide and Its Effect on Heat Resistance and Combustion Properties of Calcium Alginate
Reprinted from: *Polymers* **2019**, *11*, 1760, doi:10.3390/polym11111760 119

About the Editor

Iole Venditti

Iole Venditti is Associate Professor of Inorganic Chemistry at the Science Department of the Roma Tre University of Rome, Italy. She holds an international Ph.D. in Materials Science from Sapienza University of Rome. Her research focuses on micro and nanostructured materials, based on noble metals (gold platinum silver, copper) on polymers and hybrid systems. These materials have found application in optoelectronic devices, in drug delivery and in the detection of environmental pollutants. She has collaborated with many researchers around the world on various research projects on these topics. She is Editor for several high-level international journals, including Nanomaterials, Polymers, Chemosensors. She currently has 4 patents, has collaborated on 2 books, has published more than 100 scientific articles, with more than 3000 citations and has an h-index of 36, according to Scopus.

Preface to "Metal Nanoparticles-Polymer Hybrid Materials"

This collection focuses on a topic of great interest, hybrid metal-polymer nanomaterials. In fact, hybrid nanostructured systems based on polymers and metals are highly competitive and efficient and are starting to be present on the global market. This is why I found it stimulating to open a special issue on this topic and to fill the role of Guest Editor. The book opens with an editorial that presents the various works and their common threads. What results is characterized by a broad multidisciplinarity and boasts international contributions, with authors from different countries (Italy, Poland, USA, Korea, Czech Republic, China, Mexico,) to underline how science can always represent a bridge between peoples and cultures. All contributions represent works from leading universities and research centers around the world. I, as Guest Editor, and the MDPI staff are very pleased to offer this special issue to all readers interested in the vast world of hybrid nanomaterials.

Iole Venditti
Editor

Editorial

Metal Nanoparticles–Polymers Hybrid Materials I

Iole Venditti

Sciences Department, Roma Tre University, Via della Vasca Navale 79, 00146 Rome, Italy; iole.venditti@uniroma3.it; Tel.: +39-06-5733-3388

Citation: Venditti, I. Metal Nanoparticles–Polymers Hybrid Materials I. *Polymers* **2022**, *14*, 3117. https://doi.org/10.3390/polym14153117

Received: 11 July 2022
Accepted: 27 July 2022
Published: 30 July 2022

Publisher's Note: MDPI stays neutral with regard to jurisdictional claims in published maps and institutional affiliations.

Copyright: © 2022 by the author. Licensee MDPI, Basel, Switzerland. This article is an open access article distributed under the terms and conditions of the Creative Commons Attribution (CC BY) license (https://creativecommons.org/licenses/by/4.0/).

Important discoveries have characterized the last decade, highlighting the importance of investment in research in fields such as medicine, biology, computer science, and physics [1–9].

Hybrid nanosystems, in which different components synergistically contribute to peculiar properties, are an important example of how multidisciplinarity is now essential for developing innovative ideas and materials. In particular, the study and development of hybrid materials based on metal nanoparticles and polymers require studies on the structure–property relationships, and will allow advanced applications in sectors such as energy, optoelectronics, and environmental protection [10–16].

Therefore, this Special Issue was born, collecting contributions from different sectors such as chemistry, engineering, physics, and biology. The great goal achieved with this collection is certainly the ability to provide an updated overview of both the great opportunities and the challenges that hybrid materials based on metal–polymer nanoparticles create.

One of the key topics of the last decade concerns the use of silver nanoparticles (AgNPs) and their increasing integration into environmental applications, including the monitoring and treatment of water pollution, which is raising concerns about their environmental impact. The review by Fiorati et al. [17] analyzes the situation based on the most recent literature and proposes a design strategy that combines better performance with the absence of risks for ecosystems. The review, starting with an overview of the recent preparations of AgNPs and their uses, indicates potential solutions based on AgNPs–cellulose hybrid materials that would be efficient and eco-compatible for monitoring and treating water pollution. The development of these hybrid materials following an eco-design approach can represent a winning strategy for exploiting AgNP-based nanotechnologies for the monitoring and treatment of water pollution.

I. Jacukowicz-Sobala et al. [18] presented the preparation of composite materials containing zero-valent copper (ZVC) dispersed in a matrix of two commercially available, strongly basic anion exchangers with a macroreticular (Amberlite IRA 900Cl) and gel-like (Amberlite IRA 402OH) structure. The Cu^0 particles appeared in the resinous phase as a product of the reduction of the precursor, i.e., copper (I) oxide particles previously deposited in the two support materials. As a result, macroporous and gel-type hybrid products containing ZVC were obtained, with total copper contents of 7.7 and 5.3% by weight, respectively. X-ray diffraction and Fourier-transform infrared spectroscopy (FTIR) confirmed the successful transformation of the starting oxide particles into a metallic deposit. A scanning electron microscopy (SEM) study showed that the morphology of the deposit is mainly influenced by the type of matrix exchanger. Considering that both the reaction parameters and the operating techniques influence the size, shape, and distribution of Cu^0, as well as the porous structure of the polymer matrix, the composite materials obtained can have promising applications, including in catalytic or photocatalytic chemical reactions involving both gaseous or liquid reagents, as well as showing antibacterial activity.

G. M. Di Salvo et al. [19] studied the self-assembly of amphiphilic diblock copolymers into polymeric vesicles, commonly known as polymersomes, resulting in a versatile system for a variety of applications including drug delivery and microreactors. In this study, the incorporation of hydrophobic plasmonic nanoparticles within a polymersome membrane facilitated the light-stimulated release of vesicle encapsulants. This work sought to achieve tunable, triggered release with non-invasive, spatiotemporal control using single-pulse irradiation. Gold nanoparticles (AuNPs) were incorporated as photosensitizers into the hydrophobic membranes of micron-scale polymersomes, and the cargo-release profile was controlled by varying the pulse energy and nanoparticle concentration. The study demonstrated the ability to achieve immediate vesicle rupture as well as vesicle poration resulting in cargo diffusion. Additionally, changing the pulse duration, from femtosecond to nanosecond, provided mechanistic insight into the photothermal and photomechanical contributors that govern membrane disruption in this polymer–nanoparticle hybrid system.

W. G. Lee et al. [20] presented the preparation of porous cellulose acetate (CA) for application as a battery separator. $Cd(NO_3)_2 \cdot 4H_2O$ was utilized, with water pressure as an external physical force. When the CA was complexed with $Cd(NO_3)_2 \cdot 4H_2O$ and exposed to external water pressure, the water flux through the CA was observed, indicating the generation of pores in the polymer. Furthermore, as the hydraulic pressure increased, the water flux increased proportionally, indicating the possibility of the control of the porosity and pore size. Surprisingly, the flux value, above 250 LMH (L/m^2h), observed at the ratio of 1:0.35 (molar ratio of CA:$Cd(NO_3)_2 \cdot 4H_2O$) was higher than the fluxes of CA/other metal nitrate salt ($Ni(NO_3)_2$ and $Mg(NO_3)_2$) complexes. The higher value indicated that larger and abundant pores were generated in the cellulose acetate at the same water pressure. Thus, it could be postulated that the $Cd(NO_3)_2 \cdot 4H_2O$ salt played a role as a stronger plasticizer than the other metal nitrate salts such as $Ni(NO_3)_2$ and $Mg(NO_3)_2$. The coordinative interactions between the CA and $Cd(NO_3)_2 \cdot 4H_2O$ were investigated by IR spectroscopy. The change in ionic species in $Cd(NO_3)_2 \cdot 4H_2O$ was analyzed by Raman spectroscopy.

Tissue engineering is an area in which biomedical research is investing heavily [21,22]. Tissue engineering represents an alternative to autologous grafts. Its application requires three-dimensional scaffolds, which mimic the human body's environment. This topic is discussed by the work of J. Radwan-Pragłowska et al. [23] regarding bone-tissue engineering. In fact, bone is the second tissue to be replaced, and annually, over four million surgical treatments are performed. Bone tissue has a highly organized structure and contains mostly inorganic components. The scaffolds of the latest generation should not only be biocompatible but also promote osteoconduction. Poly (lactic acid) (PLA) nanofibers are commonly used for this purpose; however, they lack bioactivity and do not provide good cell adhesion. Chitosan is a commonly used biopolymer that positively affects osteoblasts' behavior. The aim of this study was to prepare novel hybrid 3D scaffolds containing nanohydroxyapatite capable of stimulating cell responses. Matrices were successfully obtained by PLA electrospinning and microwave-assisted chitosan crosslinking, followed by doping with three types of metallic nanoparticles (Au, Pt, and TiO_2). The products and semi-components were characterized in terms of their physicochemical properties, such as chemical structure, crystallinity, and swelling degree. The nanoparticles' and ready biomaterials' morphologies were investigated by SEM and TEM methods. Finally, the scaffolds were studied for their bioactivity on MG-63 and effects on current-stimulated biomineralization. The obtained results confirmed the preparation of tunable biomimicking matrices that may be used as a promising tool for bone-tissue engineering.

M. Wu et al. [24] introduced a three-dimensional (3D) pyramid to the polymer–plasmonic hybrid structure of polymethyl methacrylate (PMMA) composite AgNPs as a higher-quality, flexible surface-enhanced Raman scattering (SERS) substrate. The effective oscillation of light inside the pyramid valley could provide wide distributions of 3D "hotspots" in a large space. The inclined surface design of the pyramid structure could facilitate the aggregation of probe molecules, enabling the highly sensitive detection of rhodamine 6G (R6G) and crystal violet (CV). In addition, the AgNPs and PMMA composite structures provided a uniform space distribution for analyte detection in a designated hotspot zone. The incident light could penetrate the external PMMA film to trigger the localized plasmon resonance of the encapsulated AgNPs, achieving an enormous enhancement factor (~6.24×10^8). After it underwent mechanical deformation, the flexible SERS substrate still maintained high mechanical stability, which was proved by experiments and theory. For practical applications, the prepared flexible SERS substrate was adapted for the in situ Raman detection of adenosine aqueous solution and methylene-blue (MB) molecule detection on the skin of a fish, providing a direct and nondestructive active platform for detection on surfaces with any arbitrary morphology and aqueous solution.

Magnetic nanoparticles (MNPs) are largely investigated for biomedical applications. However, if not properly designed, MNPs can cause several problems, such as cytotoxicity or hemolysis. J.A. Roacho-Pérez et al. [25] compared bare magnetite nanoparticles against magnetite nanoparticles coated with a combination of polyethylene glycol 3350 (PEG 3350) and polysorbate 80 (Tween 80). The physical characteristics of the nanoparticles were evaluated. A primary culture of sheep adipose mesenchymal stem cells was developed to measure the nanoparticle cytotoxicity. A sample of erythrocytes from a healthy donor was used for the hemolysis assay. The results showed the successful obtention of magnetite nanoparticles coated with PEG 3350–Tween 80, with a spherical shape. Biological studies showed a lack of cytotoxicity at doses lower than 1000 µg/mL for mesenchymal stem cells, and a lack of hemolytic activity at doses lower than 100 µg/mL for erythrocytes.

At the end of this Special Issue, we have the work of P. Shao et al. [26] addressing the effects of cuprous nano-oxide on the flame-retardant properties of calcium alginate. Nanocuprous oxide was prepared on a sodium alginate template from which Cl^- was excluded and based on which calcium alginate/nanocuprous oxide hybrid materials were prepared by a Ca^{2+}-crosslinking and freeze-drying process. The thermal degradation and combustion behavior of the materials were studied by related characterization techniques using pure calcium alginate for comparison. The results show that the weight loss rate, heat release rate, peak heat release rate, total heat release rate, and specific extinction area of the hybrid materials were remarkably lower than those of pure calcium alginate, and the flame-retardant performance was significantly improved. In fact, nanocuprous oxide formed a dense protective layer of copper oxide, calcium carbonate, and carbon by lowering the initial degradation temperature for the polysaccharide chain during thermal degradation and catalytically dehydrating to char in the combustion process; thereby, it could isolate combustible gases, increase carbon residual rates, and notably reduce heat release and smoke evacuation.

In conclusion, as the editor of this Special Issue, I am aware that the diversity and innovation of new compounds and tools that are rapidly developing in the field of multi-disciplinary research related to nanomaterials based on noble metals cannot all be collected in a single volume. However, I am certain that this collection will contribute to the interest in the research in this area, providing our readers with a broad and updated overview of this topic.

Funding: This research received no external funding.

Acknowledgments: A special thank you goes to all the authors for submitting their studies to the present Special Issue. Moreover, the Grant of Excellence Departments, MIUR-Italy (ARTICOLO 1, COMMI 314337 LEGGE 232/2016) and Regione Lazio (Progetti Gruppi di Ricerca 2020—protocollo GeCoWEB n. A0375-2020-36521, CUP E85F21002440002) are gratefully acknowledged.

Conflicts of Interest: The author declares no conflict of interest.

References

1. Gnecco, G.; Pammolli, F.; Tuncay, B. Welfare and research and development incentive effects of uniform and differential pricing schemes. *Comput. Manag. Sci.* **2022**, *19*, 229–268. [CrossRef]
2. Piddock, L.J.V.; Paccaud, J.-P.; O'Brien, S.; Childs, M.; Malpani, R.; Balasegaram, M. A Nonprofit Drug Development Model Is Part of the Antimicrobial Resistance (AMR) Solution. *Clin. Infect. Dis. Off. Publ. Infect. Dis. Soc. Am.* **2022**, *74*, 1866–1871. [CrossRef] [PubMed]
3. Linaza, R.V.; Genovezos, C.; Garner, T.; Panford-Quainoo, E.; Roberts, A.P. Global antimicrobial stewardship and the need for pharmaceutical system strengthening for antimicrobials within a One Health approach. *Int. J. Pharm. Pract.* **2022**, *30*, 175–179. [CrossRef] [PubMed]

4. Venditti, I.; Cartoni, A.; Fontana, L.; Testa, G.; Scaramuzzo, F.A.; Faccini, R.; Terracciano, C.M.; Camillocci, E.S.; Morganti, S.; Giordano, A.; et al. Y3+ embebbed in polymeric nanoparticles: Morphology, dimension and stability of composite colloidal system. *Colloid Surf. A* **2017**, *532*, 125–131. [CrossRef]
5. Roberson, T.M. On the Social Shaping of Quantum Technologies: An Analysis of Emerging Expectations through Grant Proposals from 2002–2020. *Minerva* **2021**, *59*, 379–397. [CrossRef]
6. Cesati, M.; Mancuso, R.; Betti, E.; Caccamo, M. A memory access detection methodology for accurate workload characterization. In Proceedings of the IEEE 21st International Conference on Embedded and Real-Time Computing Systems and Applications, Hong Kong, China, 19–21 August 2015; 7299854. pp. 141–148. [CrossRef]
7. Sharma, R.; Ghosh, U.B. Cognitive Computing Driven Healthcare: A Precise Study Studies in Computational. *Intelligence* **2022**, *1024*, 259–279. [CrossRef]
8. Gioiosa, R.; Mutlu, B.; Lee, S.; Vetter, J.; Picierro, G.; Cesati, M. The minos computing library: Efficient parallel programming for extremely heterogeneous systems GPGPU 2020. In Proceedings of the 2020 General Purpose Processing Using GPU 2020, San Diego, CA, USA, 23 February 2020; pp. 1–10. [CrossRef]
9. Song, K.; An, K.; Yang, G.; Huang, J. Risk-return relationship in a complex adaptive system. *PLoS ONE* **2012**, *7*, e33588. [CrossRef]
10. D'Amato, R.; Venditti, I.; Russo, M.V.; Falconieri, M. Growth Control and Long range Self-assembly of Polymethylmethacrylate Nanospheres. *J. Appl. Polym. Sci.* **2006**, *102*, 4493–4499. [CrossRef]
11. Bonomo, M.; Naponiello, G.; Venditti, I.; Zardetto, V.; di Carlo, A.; Dini, D. Electrochemical and photoelectrochemical properties of screen-printed nickel oxide thin films obtained from precursor pastes with different compositions. *J. Electrochem. Soc.* **2017**, *64*, H137–H147. [CrossRef]
12. Shi, Q.; Qin, Z.; Xu, H.; Li, G. Heterogeneous Cross-Coupling over Gold Nanoclusters. *Nanomaterials* **2019**, *9*, 838. [CrossRef]
13. Naponiello, G.; Venditti, I.; Zardetto, V.; Saccone, D.; di Carlo, A.; Fratoddi, I.; Barolo, C.; Dini, D. Photoelectrochemical characterization of squaraine-sensitized nickel oxide cathodes deposited via screen-printing for p-type dye-sensitized solar cells. *Appl. Surf. Sci.* **2015**, *56*, 911–920. [CrossRef]
14. Schutzmann, S.; Venditti, I.; Prosposito, P.; Casalboni, M.; Russo, M.V. High-energy angle resolved reflection spectroscopy on three-dimensional photonic crystals of self-organized polymeric nanospheres. *Opt. Express* **2008**, *16*, 897–907. [CrossRef]
15. Sayson, L.V.A.; Regulacio, M.D. Rational Design and Synthesis of Ag–Cu$_2$O Nanocomposites for SERS Detection, Catalysis, and Antibacterial Applications. *ChemNanoMat* **2022**, *8*, e202200052. [CrossRef]
16. Qamar, S.A.; Qamar, M.; Basharat, A.; Bilal, M.; Cheng, H.; Iqbal, H.M. Alginate-based nano-adsorbent materials—Bioinspired solution to mitigate hazardous environmental pollutants. *Chemosphere* **2022**, *288*, 132618. [CrossRef]
17. Fiorati, A.; Bellingeri, A.; Punta, C.; Corsi, I.; Venditti, I. Silver Nanoparticles for Water Pollution Monitoring and Treatments: Ecosafety Challenge and Cellulose-Based Hybrids Solution. *Polymers* **2020**, *12*, 1635. [CrossRef]
18. Jacukowicz-Sobala, I.; Stanisławska, E.; Baszczuk, A.; Jasiorski, M.; Kociołek-Balawejder, E. Size-Controlled Transformation of Cu$_2$O into Zero Valent Copper within the Matrix of Anion Exchangers via Green Chemical Reduction. *Polymers* **2020**, *12*, 2629. [CrossRef]
19. di Salvo, G.M.; Robinson, A.R.; Aly, M.S.; Hoglund, E.R.; O'Malley, S.M.; Griepenburg, J.C. Polymersome Poration and Rupture Mediated by Plasmonic Nanoparticles in Response to Single-Pulse Irradiation. *Polymers* **2020**, *12*, 2381. [CrossRef]
20. Lee, W.G.; Cho, Y.; Kang, S.W. Effect of Ionic Radius in Metal Nitrate on Pore Generation of Cellulose Acetate in Polymer Nanocomposite. *Polymers* **2020**, *12*, 981. [CrossRef]
21. Ashammakhi, N.; GhavamiNejad, A.; Tutar, R.; Fricker, A.; Roy, I.; Chatzistavrou, X.; Apu, E.H.; Nguyen, K.-L.; Ahsan, T.; Pountos, I.; et al. Highlights on Advancing Frontiers in Tissue Engineering. *Tissue Eng. Part B Rev.* **2022**, *28*, 633–6641. [CrossRef]
22. Fardjahromi, M.A.; Nazari, H.; Tafti, S.M.A.; Razmjou, A.; Mukhopadhyay, S.; Warkiani, M.E. Metal-organic framework-based nanomaterials for bone tissue engineering and wound healing. *Mater. Today Chem.* **2022**, *23*, 100670. [CrossRef]
23. Radwan-Pragłowska, J.; Janus, Ł.; Piątkowski, M.; Bogdał, D.; Matysek, D. 3D Hierarchical, Nanostructured Chitosan/PLA/HA Scaffolds Doped with TiO2/Au/Pt NPs with Tunable Properties for Guided Bone Tissue Engineering. *Polymers* **2020**, *12*, 792. [CrossRef] [PubMed]
24. Wu, M.; Zhang, C.; Ji, Y.; Tian, Y.; Wei, H.; Li, C.; Li, Z.; Zhu, T.; Sun, Q.; Man, B.; et al. 3D Ultrasensitive Polymers-Plasmonic Hybrid Flexible Platform for In-Situ Detection. *Polymers* **2020**, *12*, 392. [CrossRef] [PubMed]
25. Roacho-Pérez, J.A.; Ruiz-Hernandez, F.G.; Chapa-Gonzalez, C.; Martínez-Rodríguez, H.G.; Flores-Urquizo, I.A.; Pedroza-Montoya, F.E.; Garza-Treviño, E.N.; Bautista-Villareal, M.; García-Casillas, P.E.; Sánchez-Domínguez, C.N. Magnetite Nanoparticles Coated with PEG 3350-Tween 80: In Vitro Characterization Using Primary Cell Cultures. *Polymers* **2020**, *12*, 300. [CrossRef] [PubMed]
26. Shao, P.; Xu, P.; Zhang, L.; Xue, Y.; Zhao, X.; Li, Z.; Li, Q. Non-Chloride in Situ Preparation of Nano-Cuprous Oxide and Its Effect on Heat Resistance and Combustion Properties of Calcium Alginate. *Polymers* **2019**, *11*, 1760. [CrossRef]

Review

Silver Nanoparticles for Water Pollution Monitoring and Treatments: Ecosafety Challenge and Cellulose-Based Hybrids Solution

Andrea Fiorati [1], Arianna Bellingeri [2], Carlo Punta [1], Ilaria Corsi [2] and Iole Venditti [3,*]

[1] Department of Chemistry, Materials, and Chemical Engineering "G. Natta" and INSTM Local Unit, Politecnico di Milano, Piazza Leonardo da Vinci 32, I-20133 Milano, Italy; andrea.fiorati@polimi.it (A.F.); carlo.punta@polimi.it (C.P.)
[2] Department of Physical, Earth and Environmental Sciences and INSTM Local Unit, University of Siena, 53100 Siena, Italy; arianna.bellingeri@student.unisi.it (A.B.); ilaria.corsi@unisi.it (I.C.)
[3] Department of Sciences, Roma Tre University of Rome, via della Vasca Navale 79, 00146 Rome, Italy
* Correspondence: iole.venditti@uniroma3.it; Tel.: +39-06-5733-3388

Received: 7 July 2020; Accepted: 20 July 2020; Published: 23 July 2020

Abstract: Silver nanoparticles (AgNPs) are widely used as engineered nanomaterials (ENMs) in many advanced nanotechnologies, due to their versatile, easy and cheap preparations combined with peculiar chemical-physical properties. Their increased production and integration in environmental applications including water treatment raise concerns for their impact on humans and the environment. An eco-design strategy that makes it possible to combine the best material performances with no risk for the natural ecosystems and living beings has been recently proposed. This review envisages potential hybrid solutions of AgNPs for water pollution monitoring and remediation to satisfy their successful, environmentally safe (ecosafe) application. Being extremely efficient in pollutants sensing and degradation, their ecosafe application can be achieved in combination with polymeric-based materials, especially with cellulose, by following an eco-design approach. In fact, (AgNPs)–cellulose hybrids have the double advantage of being easily produced using recycled material, with low costs and possible reuse, and of being ecosafe, if properly designed. An updated view of the use and prospects of these advanced hybrids AgNP-based materials is provided, which will surely speed their environmental application with consequent significant economic and environmental impact.

Keywords: silver nanoparticles; nanocellulose; engineered nanomaterials; water monitoring; water treatment; ecosafety; ecotoxicology; eco-design

1. Introduction

Engineered nanomaterials (ENMs) are typically defined as materials smaller than 100 nm in at least one dimension, and they have specific surface functionalizations and size-dependent properties, such as high reactivity and large surface-to-volume ratio, that satisfy their wide range of applications, including in sensing [1–5], optics [6–10], energy [11–15], catalysis [16–19], biotechnology [20–24] and so on.

Among others, silver nanoparticles (AgNPs) have aroused great interest due to their low cost, synthetic versatility, and chemical-physical properties. For this reason, they are already widely used and present in various commercial products, often as hybrid compounds, that can be responsive to external stimuli [25–30]. In fact, although not fully addressed in terms of environmental impact, AgNPs have great marketing value, and their production is expected to reach approximately 800 t by 2025 [17,31,32]. The use of AgNPs is often linked to their antibacterial properties, but in this review, we do not discuss this aspect, for which we refer to the extensive and comprehensive literature already published [29,33,34]. Rather, more recent contributions have revealed the successful use of AgNPs as plasmonic sensors for water pollutants such as heavy metals and organic compounds, and as suitable photocatalysts for promoting the oxidative degradation of the latter, above all dyes and pesticides, which enlarges the field of environmental applications [35–38].

This review envisages potential solutions of AgNPs in water pollution monitoring and remediation, with particular emphasis on their environmentally safe application. It starts from an (i) updated overview on the incredible features of AgNPs and their easy, cheap and versatile synthesis, moves to (ii) the ecosafety concept by highlighting the relevance of assessing their environmental impact in terms of toxicity for aquatic species achieved by an eco-design approach, and (iii) provides potential advanced hybrids solutions such as cellulose–AgNP composites, which allow synergistic and eco-friendly actions. This review was conceived with the awareness that only a multidisciplinary and integrated approach among biological, chemical, and engineering visions allows a real deepening of the topic of new hybrid ecosafe solutions for water pollution monitoring and treatment and an enrichment that goes beyond the simple collection and proposal of recent articles.

2. AgNPs Preparation and Use for Water Pollution Monitoring and Treatment

2.1. AgNP Synthesis and Characterizations

The need to improve the performance of sensors in terms of sensitivity, selectivity, reusability and eco-sustainability has led to strong research and development in the field of ENMs [39–42]. In fact, thanks to the high surface-to-volume ratio, they allow a better interaction with analytes. Furthermore, the possibility of modulating size, shape and functionalization allows their chemical-physical properties to be controlled, such as optical or assembly properties, in order to adapt them to specific needs [43–48].

AgNPs have been widely used in sensing applications due to their surface properties, such as local surface plasmon resonance (LSPR). LSPR is known to be due to electrons on the surface of noble metal NPs that interact with electromagnetic radiation and produce strong extinction and dispersion spectra, at about 400 nm in the visible spectrum range. This property is incredibly useful for detection. In fact, when an analyte arrives on the surface of AgNPs, the LSPR band can be modified, and very often a change in color can also occur [35,49–55]. Moreover, AgNPs offer many advantages over other metal NPs (Au, Cu, Li, and Al NPs) due to the possibility of showing LSPR in the visible (vis) and near-infrared regions (NIR) in the range 300–1200 nm [56–58]. Ag also has the highest electrical and thermal conductivity among all metals, making it an ideal component for electrical interconnection. When exposed to air, Ag is not oxidized, but forms a silver sulfide film on its surface, which should be more or less transparent to visible light [59,60]. Ag is relatively cheap among metals that support plasmons, and together with its ease of nanofabrication, this determines its usefulness as a metal for plasmonic applications, especially on a large scale.

The LSPR makes AgNPs very interesting for the design of photocatalytic materials active under sunlight radiation. Indeed, AgNPs are already used as efficient photocatalysts for a variety of reactions, including the mineralization of organic pollutants [38,61]. The great advantage of photocatalysis consists of the direct conversion of light energy into chemical energy, thus reducing energy consumption and environmental pollution, in accordance with general rules of sustainable chemistry and green organic synthesis. For these applications, a key point consists of the selection of the proper support for the immobilization of the active catalyst particles. So far, several solutions have been studied, including polymers, cellulose, carbon materials, mesoporous materials, and so on [62,63].

On the other hand, some issues remain. The main concern is related to colloidal instability [33]. Colloidal stability depends on several experimental factors including the type of capping agent and surrounding environmental conditions, such as pH and ionic strength. Many capping agents (fo example, thiols, polymers, and surfactants) have been investigated to improve AgNPs suspension stability, preventing their aggregation through electrostatic repulsion, steric hindrance or both. Often the prevalent capping agent is citrate, but citrate-stabilized AgNPs aggregate quickly in standard biological media, such as natural sea water (pH = 7.0), phosphate-buffered saline (PBS pH = 7.2–7.4), and acetate buffer (pH = 5.6).

Many methods for synthesizing stable AgNPs by chemical, physical and biological processes, using both bottom-up and top-down approaches, have been investigated [64–68]. The well-assessed bottom up approach is based on a wet reduction of Ag^+ ions in the presence of capping agents. Generally, hydrophilic AgNPs can be synthesized starting from water solution of $AgNO_3$ in presence of ligand molecule, and adding a reducing agent, such as sodium borohydride or formic acid [24,68–71].

By choosing the appropriate method and experimental parameters for the synthesis it is possible to control shapes and obtain spheres (AgNPhs), cubes (AgNCs), stars (AgNSs) and rods (AgNRs) [72–100]. Some examples with TEM images showing AgNPs with different geometries and shapes are presented in Figure 1 [68,81,88].

Figure 1. Synthesizing Ag nanostructures: (**A**) spheres, (**B**) cubes, (**C**) stars and (**D**) rods. Reprinted with permission from [68,81,88].

Among other shapes, the spherical one is the most symmetrical and is the most used because of the easy synthetic control of reproducibility and monodispersity, both being key aspects for ensuring the repeatability of experiments and results. The AgNPhs are synthesized by Turchevich and Shiffrin Brust methods and their modifications, producing the typical LSPR in the range 370–470 nm [26,54,72–77]. The well-assessed synthetic studies make it possible to also introduce these silver nanomaterials in nanocomposites, finding application above all in biotechnology and as sensors [55,74,78,79].

AgNCs are very promising anisotropic materials. The presence of corners and edges in nanocubes allows an increase in the signal due to the increase in local field. Despite these better properties, the synthesis of AgNCs is still considered difficult, especially as regards the monodispersity quality and reproducibility. In fact, in AgNCs there is material confined by six 100 closed planes, which require precise growth settings for their formation. The experimental parameters that strongly influence these ENMs are the precursor concentrations, the mixing conditions, the temperature and the reaction time. To solve these problems, different methods and techniques have been studied and among them, the microfluidic platforms showed promising perspectives [80–85].

Star colloidal suspensions of silver (AgNSs) were prepared by chemical reduction of Ag^+ in two steps and using different reducing and capping agents in each step. The number of star arms varies, with eight generally being the average number. Sometimes these arms can be branched. The AgNSs can have an average diameter from 70 to 700 nm. In general, the tips of the arms display a low sharpness. The extinction spectra the bands are displayed at about 370 nm, and there are extinction background bands at longer wavelengths, where weak maxima are distinguished in the range 650–750. The large extinction background can be attributed to the absorption and scattering emissions produced by the different morphologies of all the existing NPs, integrated by AgNSs bearing different numbers of arms and having different tip sharpness. The mixture of all these factors leads to very different LSPR in the suspension, since a wide range of wavelengths in the visible and near-IR ranges can be covered with various AgNSs shapes and dimensions [86–93].

AgNRs have an anisotropic shape that produces two plasmon bands: the transverse plasmon band due to an electron oscillation along the short axis of the rod, at around 520–550 nm, and the longitudinal plasmon band, in the range 800–1200 nm. Therefore, these ENMs are active in the NIR attracting interest for biomedical application as therapeutic and imaging agents [94–96]. Moreover, their aspect ratio (length divided by width) and surface functionalization are easily adjustable by synthetic parameters. There are several AgNR synthesis methods, but the most used are based on seed-based colloidal growth methods, in two steps: in the first step, nucleation is obtained separately by producing seeds which in the second step are added to the growth solution of the nanorods [97,98]. Another widespread technique is based on the oblique angle deposition method (OAD). OAD is a physical vapor deposition method in which the vapor is incident with a large angle ($\theta > 70°$) with respect to the normal surface of substrates, and nanorods or wires are usually formed. By using OAD it is possible to obtain narrower size and more uniform distribution of AgNRs [99,100].

Furthermore, the functionalization and engineering of the Ag surface can be used to guide the aggregation–disaggregation phenomena of ENMs under specific conditions (pH, temperature change or presence of different analytes), leading LSPR band changes. This makes it possible to have a responsive material, capable of interacting with the environment, when properly functionalized. The main strategies for surface functionalization of AgNPs are schematized in Figure 2.

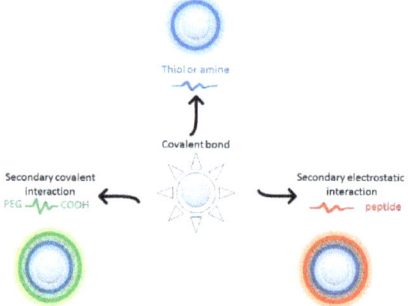

Figure 2. Scheme of main functionalization strategies for AgNPs.

Direct functionalization takes place during synthesis by choosing molecules that stabilize the first clusters of metal particles that are formed by the reduction of Ag(I), generally by sodium borohydride or analogous reducing agents. Usually, these capping molecules exhibit thiol or amine functionalities in which S or N bind covalently to the silver surface. The steric hindrance and any charges present on the cap molecules can induce specific interaction behaviors between particles. After the first functionalization in synthesis, it is possible to resort to a secondary functionalization that engages a second layer of molecules. Electrostatic interactions or covalent bond formation between molecules of the first and second covering layers of AgNPs can be exploited. As for electrostatic interactions, in several cases, the second layer is formed by a peptide, while for the formation of covalent bonds polyethylenglycol (PEG) is often used as ideal coating agent.

2.2. AgNPs for Water Pollution Monitoring and Treatment

The properties and versatility of AgNPs have led them to be widely used in sensor systems. In fact, by changing the type of surface functionalization and by choosing a specific ligand, it is possible to make particles selective to a particular analyte, and by optimizing the degree of surface functionalization, their sensitivity can be improved. The main chemical molecules and capping agents used for direct functionalization of AgNPs during the synthesis of different morphologies are reported in Table 1, together with dimension and applications for water pollution monitoring [26,54,72–74,76–88,101–108].

Table 1. Main morphologies reported with dimension, surface functionalities and application for water pollution monitoring and treatments.

Shape	Dimension (nm)	Surface Functionalization	Detection and Monitoring of Pollutant	Ref.
Spheres				
	10–15	chalcone carboxylic acid	Cd(II)	[72]
	5–8	sodium 3-mercapto-1-propanesulfonate	Co (II); Ni (II)	[26]
	-	methyl cellulose	Cu (II)	[74]
	10–20	Riboflavin	Hg (II)	[79]
	5–8	Citrate/L-cysteine	Hg (II)	[54]
	10–15	ciclodextrin	Hg (II)	[107]
	6	Thiol terminated chitosan	Hg (II)	[108]
	20	polyvinyl alcohol	Pb(II)	[73]
	9–10	gluconate	Pb(II)	[76]
	2	polyethyleneimine (PEI)	p-nitrophenol	[101]
	5–10	thioglycolic acid	6-benzylaminopurine	[77]
	-	citrate+hexapeptide	Malathion [1]	[78]
	8–10	cyclen dithiocarbamate-	Thiram; paraquat [1]	[102]
Cubes				
	100	Poly(vinylpyrrolidone)	Paraoxon; thiram [1]	[80]
	200	glycolaldeyde	-	[81]
	90–100	cetyltrimethylammoniumcloride	-	[82]
	40–80	Poly(vinylpyrrolidone)	-	[83]
	200–300	polyaniline	hemoglobin	[84]
	60–100	hexamine	Bis phenol	[85]
Stars				
			glucose	[87]
	50–150	Lauryl sulfobetaine	Melamine	[86]
	180–250	Citrate/hydroxylamine	Congo Red	[88]
Rods				
	150–250	-	antibiotic	[103]
	-	Cy5-ssDNA	Hg(II)	[104]
	30–200	-	Polyclorinated biphenyls	[105]
	10–20	poly(ethylene glycol)	mitoxantrone	[106]

[1] Pesticides.

Manivannan et al. [107] developed AgNPhs embedded in an amine functionalized silicate sol–gel matrix, prepared by using different combination of silicate, surfactant and cyclodextrin. The authors observed the blue shift up or quenching of SPR band due to the formation of anisotropic Ag amalgam crystals. The selective sensing of Hg(II) ions by the AgNPh-based sensors in the presence of 500 mM of other environmentally relevant metal ions was verified using spectral and colorimetric methods. Sharma et al. [108] investigated the same behavior for their AgNPhs and developed a simple, label-free, cost-effective, portable, selective, and sensitive colorimetric sensor based on thiol-modified chitosan AgNPhs for the real-time detection of toxic Hg(II) ions in water. In this work, the authors observed an LSPR blue shift in the UV-vis spectra of the solution of Md-Ch-AgNPhs with the addition of the Hg^{2+}. This change in the SPR is due to the redox interaction between the Hg(II) ions and AgNPhs, ascribable to the difference in the redox potentials of Hg(II)/Hg couple (0.85 V) and Ag(I)/Ag couple (0.8 V). AgNCs were prepared by Wang et al. using sulfide-mediated polyol method [80]. These AgNCs are active substrates for SERS and allowed the detection of the pesticides paraoxon and thiram. Shkilnyy et al. [106] synthetized AgNRs coated with poly(ethylene glycol) (PEG) covalently attached to their surface. Due to steric repulsion between polymer-modified surfaces, the stability of the nanoparticle suspension was preserved even at high ionic strength (0.1 M NaCl). At the same time, the PEG coating remains sufficiently permeable, allowing surface-enhanced Raman scattering (SERS) from micromolar concentrations of small molecules such as the anticancer drug mitoxantrone (MTX).

Another important feature of AgNPs is their potential as photocatalysts.

Generally, Ag-based photocatalysts can be classified into two categories: (i) plasmonic photocatalysts based on LSPR effect of AgNPs and (ii) semiconductor photocatalysts based on bandgap excitation of Ag-containing compounds. Upon light irradiation, plasmonic excitation in AgNPs yields hot electrons, while bandgap excitation in Ag-containing semiconductors generates electron hole pairs, which undergo charge separation and transfer, and finally participate in the catalytic reactions. Recent studies have shown that Ag-based visible light responsive photocatalysts could be highly effective for organic pollutant decomposition, bacteria destruction, water reduction and oxidation, and selective organic transformation. In heterogeneous catalysis, free-standing metal NPs usually suffer from low stability due to aggregation. To circumvent this issue, anchoring plasmonic metal photocatalysts on the support materials could enable them to be highly dispersed and easily recycled.

Some examples of Ag ENMs used for polluted water treatments are reported in Table 2, pointing out AgNP size and supporting the existence of synergic effects in some cases [61,75,109–112].

Table 2. AgNP-based hybrid systems used for water pollution treatments.

AgNPs Size (nm)	Support	Treatment	Ref.
5–10	photocrosslinked matrix	nitroderivates	[61]
10–20	$BiVO_4$	crystal violet; Rhodamine B	[109]
10–30	sulfonated graphene/TiO_2	Rhodamine B; Methyl Orange; 4-nitrophenol	[110]
10		Methylene blue	[75]
5–10	TiO_2	Methyl Orange	[111]
100	cellulose nanofibrils	Rhodamine B;	[112]

A large number of innovative hybrid materials with a synergetic or complementary behavior were obtained using Ag-based photocatalysts. Ullah et al. [109] prepared bismuth vanadate ($BiVO_4$, BV) and Ag/AgO_2 hybrid nanomaterials with a significant increase (around 28 times) in photoactivity in comparison to pristine BV, as measured by photodegradation of crystal violet and Rhodamine B employing commercial low cost blue LEDs or natural sunlight as photoexcitation sources. The enhanced photoactivity of BV/Ag/AgO_2 photocatalysts is attributed to the improved adsorption of dyes and extended absorption of visible light by the photocatalysts, and better charge transfer kinetics. Melinte et al. [61] prepared several photocatalysts based on Ag, Au or Au-Ag nanoparticles supported on photocrosslinked organic and these hybrid systems allowed the photocatalytic degradation of

4-nitroaniline. Roy et al. [75] studied the photocatalytic degradation of methylene blue dye in presence of biogenic AgNPs synthesized using yeast (*Saccharomyces cerevisiae*) extract.

The treatment of water with silver ENMs has also been performed in consideration of their antibacterial action. In fact, AgNPs have wide applications in water disinfection thanks to their well-known antimicrobial activity. Various mechanisms have been proposed in the literature to explain the antimicrobial activity of AgNP which can be traced back to three actions: (1) alteration of membrane properties; (2) damage to DNA/RNA and/or proteins; or (3) release of Ag (I) in the cell cytoplasm. Although in this review the antimicrobial and antibacterial properties of AgNPs are not the focus, how these properties can influence and cause effects on the environment, producing important environmental impacts and causing rebound to their safe application will be discussed in the next paragraph. Today, these are a new ecological challenge.

3. Ecosafety Challenges

3.1. Environmental Safety of ENMs

The ecotoxicological implications of ENMs for aquatic and terrestrial organisms have been widely documented and mainly attributed to the peculiar nanoscale properties (i.e., size, shape, surface charges) and their transformation once released in natural ecosystems [113].

A large number of bench-scale studies have clearly identified in the nanoscale dimension and surface chemistry the main drivers of cellular uptake by which ENMs can be easily internalized by the cell through different processes (i.e., phagocytosis, endocytosis, direct trans-membrane transport) and exert their toxic action [114,115].

A slight change in NP size (agglomeration or aggregation) and/or surface chemistry (i.e., interaction with ionic species and colloidal particles) can significantly affect behavior and exposure dynamics towards living organisms [116–118]. Processes such as homoaggregation (i.e., aggregation of ENMs of the same nature) and heteroaggregation (i.e., the aggregation between non-homologous particles), which are dependent on the chemistry of the receiving environmental media (i.e., soil, freshwater and saline) and on the NP functional coatings, may limit their ability to be internalized by the cells as well as their dissolution capabilities [119,120]. On the other hand, processes such as resuspension and disaggregation, luckily occurring in natural water bodies (river, estuaries and oceans) could still make bioavailable a consistent fraction of ENMs to aquatic organisms along the water column or in sediments [121]. Similarly, processes such as oxidation, sulfidation, chlorination and dissolution, more frequent for metal-based ENMs, could significantly affect their environmental behavior and toxicity [29].

AgNPs are the most commonly and widely used ENMs in consumer products, mainly due to their biocidal properties which makes them highly efficient anti-microbial agents [1,2,17,122–125]. A study conducted on two German wastewater treatment plants demonstrated the removal of AgNPs up to 96.4%, with residual concentration ranging from 0.7 to 11.1 ng/L in effluents. However, based on such data, despite the high removal efficiency, the authors estimated a total release in the environment of 33 Kg AgNPs/year for the whole country [126]. For these reasons, AgNPs have stimulated considerable attention in terms of potential environmental risks consequent to their production, usage, disposal and application [127–130].

3.2. AgNP Toxicity to Aquatic Biota

The antimicrobial activity of AgNPs is closely linked to the release of Ag^+ ions [131], which are recognized as one of the most toxic metal ions in the aquatic environment [132]. The use of AgNPs, compared to $AgNO_3$, gives a much effective and longer-term antimicrobial activity, due to the continuous and prolonged release of Ag^+ ions, close to the target organism [133,134]. This feature confers a high potential hazard to the environmental release of AgNPs, since their biocidal properties are not only restricted to microorganisms. In fact, AgNPs are largely documented to be toxic to aquatic biota [30,135] including freshwater [18,39,136] and marine species [34,136–140] (Figure 3).

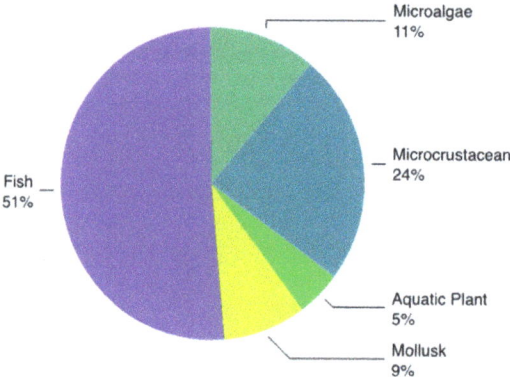

Figure 3. Schematic representation of the distribution of taxa investigated in aquatic ecotoxicity studies with AgNPs up to 2019 (Scopus source). Total number of studies was 282.

As for microorganism, AgNPs' ecotoxicity to non-target species is mainly imputable to dissolution, even if some studies report that toxicity seems to not be fully explained by dissolved Ag and that a nano-specific toxicity could be involved [128,129]. Additionally, additional effects can be observed as particle size decrease [141,142], even though there is no general consensus on the contribution of each factor [143]. Smaller particles could influence toxicity by providing a larger surface area for the particle's dissolution [144] or allow cellular internalization [145], resulting in interaction with molecular mechanisms or intracellular dissolution [146].

Numerous studies investigating the toxic effect of AgNPs to non-target organisms are lacking information about particles behavior and dissolution in the media, which is a key element for understanding the mechanism of toxicity and thus their environmental safety. The presence in environmental media of oxidizing species, chlorines or sulfur, could favor the release of Ag^+ ions and therefore significantly affect toxic action not only towards microbes but in general to all potentially exposed organisms [127].

3.3. Role of the Surface Coating in AgNPs Ecotoxicity

Among the intrinsic factors influencing the dissolution of the particles, the type of surface coating plays a crucial role, by also influencing particles aggregation state and overall stability in the media [147]. Experimental evidence shows that, in the same conditions, AgNPs with different types of coating result in different toxicities [18,129,145,148]. A safe design strategy was proposed by Pang et al. [130], who demonstrated, both in vitro (Hepa1c1c7) and in vivo (mice), that biodistribution, intracellular localization and toxicity of AgNPs were significantly affected by NP surface coatings and in particular by the presence of positive surface charges (BPEI AgNPs > Citrate AgNPs = PVP AgNPs > PEG AgNPs).

Some coating agents can prevent the dissolution of the particles, either by closely bind AgNPs surface and preventing Ag^+ ions to solubilize or by excluding oxidizing agents to come in contact with the particle's surface, preventing the particle's oxidation and consequent dissolution [70]. That is the case of molecules with a high reduced sulfur content that bind metal ions with high affinity [149]. The toxicity of AgNPs for both microbes and non-target species has been shown to decrease following the addition of molecules rich in thiol groups in solution, such as natural organic matter or cysteine [18,150–152].

Several studies have assessed the toxicity of sulfur functionalized AgNPs, and always reported a reduction in toxicity compared to pristine AgNPs. Levard et al. [153] reported that, as the level of sulfidation of AgNPs increased, the toxicity to four different model organisms (fish embryos, worm and aquatic plant), as well as the dissolution of the particles, decreased. Another study reported that AgNPs functionalized either with cysteine or glutathione showed a lower toxicity to the microcrustacean *Daphnia magna* compared to reported literature data on differently coated AgNPs [154].

3.4. Ecosafe by Design Approach

Ecosafety has begun to be included in a risk assessment framework to support policy-makers, legislators and stakeholders with the aim of limiting any risk associated with their use and application [29,155]. To achieve environmental safety, the new strategy widely adopted in nanomedicine to design safer materials has been recognized to be successful in order to disregard those undesirable properties which can be hazardous for humans and the environment [32]. A similar concept was recently adopted in the design of ENMs for environmental application as for instance in pollution remediation (nanoremediation), in which environmental safety and efficacy towards removal of pollutants from an environmental setting will be incorporated in the material design process (eco-design) [28,32,156,157]. Removal efficacy and ecotoxicity are tested in parallel in order to obtain best adsorption performances with no risk for the exposed organisms (Figure 4).

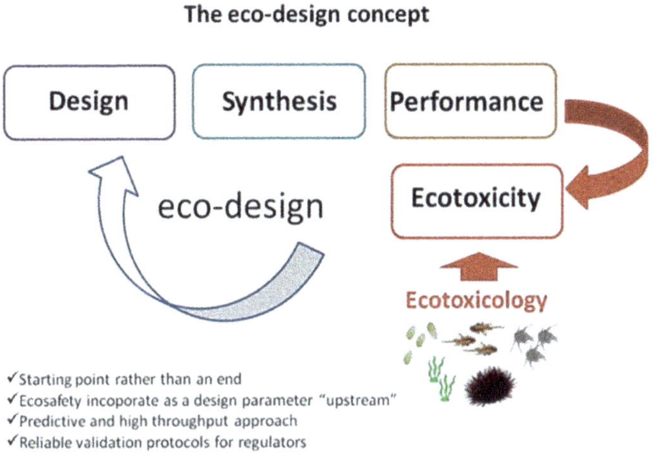

Figure 4. The eco-design concept based on ecotoxicological risk assessment as a tool for eco-design achievement [158].

By reducing nanomaterial uncertainties and environmental and human risks, an eco-design strategy could support regulatory requirements, boost circular economy and act as a further driver for market development of the nanoremediation sector.

As an example, in line with this strategy, a recent contribution in the field was made by selecting ecosafe batches of cellulose-based nanosponges (CNS) for heavy metal removal from seawater obtained by an eco-design strategy [147]. A trial-and-error-like process in which direct effects on aquatic organisms of CNS, evaluated by an ecotoxicological approach, aids the formulation of CNS in a stepwise fashion (i.e., by testing single components and synthesized materials at the final stage) along with the synthetic procedure and it has been proven to be successful in selecting the most environmental safe CNS formulation [156,158].

By testing two aquatic trophic levels, planktonic and filter-feeder benthic species, as potential biological targets of CNS, the eco-design strategy allows to select those ecologically safer material properties to avoid any potential ecological risks consequent to their environmental application. A more recent contribution clearly demonstrated that effect-based tools as biological responses in exposed organisms can be used to validate adsorption efficacy of CNS towards Zn removal from seawater and prevent potential ecological risks due to their application [159,160].

Furthermore, the same eco-design strategy was recently applied with a newly synthesized batch of AgNPs functionalized with citrate and L-cysteine (AgNPs@Cit/L-cys), and thus was able to selectively adsorb mercury (Hg^{2+}) from water. This peculiar functionalization apparently prevents the release of Ag^+ confirmed by the absence of toxicity towards freshwater and marine microalgae (*Raphidocelis subcapitata* and *Phaeodactylum tricornutum*) (Figure 5).

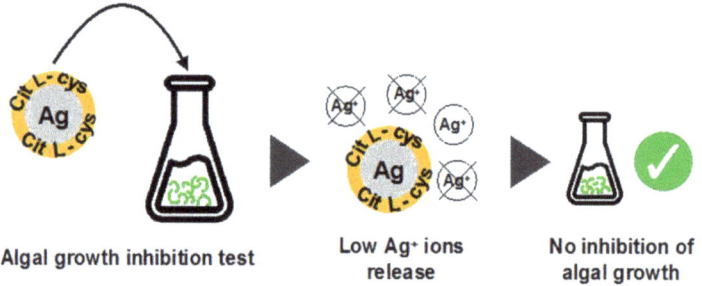

Figure 5. Testing of AgNPcitLcys for algal toxicity resulted in low release of silver ions and no inhibition of algal growth, from both fresh- and marine waters [54].

The high adsorption capability towards Hg^{2+} coupled with no risks for aquatic species clearly support their application for in situ remediation. For more details refer to Prosposito et al. [54]

Therefore, a more comprehensive understanding of the role of AgNPs surface coatings, not only in driving their behavior in environmental settings but also their ability to be taken up by aquatic organisms, is becoming mandatory for supporting ecosafety of ENM for environmental application. In this regard, the need to predict realistic exposure scenarios as for instance the in situ and ex situ applications stimulates the adoption of an effect-based approach which will mimic environmental exposure conditions [54,159].

4. Cellulose Doped with AgNPs: A Synergic Solution

4.1. A Sustainable Solution for AgNP Immobilization

A valuable route to overcoming the concerns related to the direct use of AgNPs, due to the potential toxicity of these systems, consists of their immobilization on proper supports. This approach, which would allow the limiting of NP migration and even the improvement of remediation action by preventing their agglomeration due to surface energy, takes inspiration from technologies developed to transfer to textiles the most investigated property of AgNPs, namely the antimicrobial activity. For this reason, biopolymers have been widely considered as ideal substrates for AgNP binding [158,161].

While the first applications of this approach were in the field of textile materials for medical, health care, and hygienic uses, researchers envisioned the possibility of also applying AgNPs/biopolymer composites for the safe sanitization of drinking water [162,163].

In consideration of the previously discussed adsorption/degradation action of silver nanoparticles, this technology has more recently been extended to the design of devices, filters, and membranes for wastewater decontamination. The knowledge achieved in the field of textile functionalization fits well with this specific target in terms of materials of choice for supporting the nanoparticles. In fact, biopolymers such as cellulose, starch-derived dextrin, and chitosan are becoming more and more attractive as ideal building blocks for the production of smart materials for water treatment [164]. Polysaccharides allow the combination of their renewable and biodegradable nature with a negligible toxicity, meeting the ecosafety premise necessary to follow the eco-design of sustainable solutions for the environmental monitoring and remediation, as discussed in Section 3 [32].

Cellulose is the most abundant biopolymer on Earth, as it can be extracted from a wide range of renewable sources, such as cotton, wood, and disposed biomass, including agricultural waste and recycled paper (Figure 6a) [156]. Cellulose fibers present a high-mass molecular structure, formed by the repetition of a huge number of β-D-glucopyranose monomers linked together by β-1,4 glycosidic bonds [165]. This structure favors the synthesis and binding of nanoparticles, acting as stabilizing and capping agent.

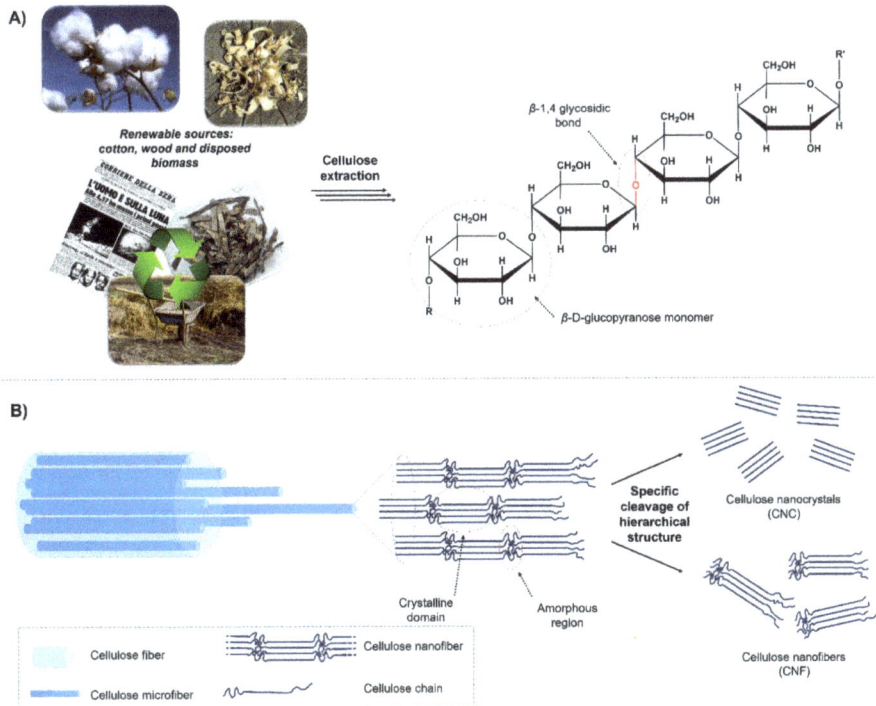

Figure 6. (**A**) Cellulose sources and polymer structure; (**B**) hierarchical structure of cellulose nanofibers.

Moreover, it is possible to cleave the hierarchical structure of cellulose fibers by following different mechanical or chemical treatments, producing nanocellulose (NC) in the form of nanocrystals (CNC) or nanofibers (CNF), the latter presenting crystalline domains alternating with amorphous regions (Figure 6b) [166–168]. Moving to a nanoscale dimension it is possible to increase significantly the surface area and the biopolymer reactivity, opening the way to a selective functionalization of NC, in order to introduce additional moieties for enforcing nanoparticle immobilization. The introduction of new much more environmentally and economically sustainable protocols, such as the 2,2,6,6-tetramethylpiperidine 1-oxyl (TEMPO)-mediated oxidation [169,170] and the enzymatic (endoglucanase) pretreatment [170–172], favored the scaling up of NC production up to the order of hundreds of kilograms per day (http://www.tappinano.org, 2019).

Finally, cellulose- and even more nanocellulose-based materials have been reported to be able to efficiently adsorb a wide range of transition metal ions and organic dyes and pollutants [158,173–175]. As a consequence, it is possible to combine, in the same composite, the high adsorption efficiency of NC with the catalytic or photocatalytic action of AgNPs for promoting the degradation of contaminants in water.

4.2. Cellulose Doping with Pre-Formed AgNPs

The simplest approach for the synthesis of AgNPs/cellulose composites consists of the use of commercially available NPs or their pre-formation in a separate batch. In this case, cellulose doping occurs by contacting the NP aqueous dispersion with the polymeric support. Sehaqui and coworkers reported the synthesis of CNF-based filters and the capture of a wide range of nanoparticles, including silver ones, by simple filtration [176]. CNF were first functionalized in part with quaternary ammonium moieties using 2,3-epoxypropyl trimethyl ammonium chloride, and in part with carboxylic groups by means of succinic anhydride. The filter was prepared by freeze-drying a homogeneous dispersion of the two typologies of CNF, providing a 2D sheet-like microporous structure, in line to what reported for similar systems [176–178].

Gold and silver nanoparticles were fixed thanks to the electrostatic interaction between the positive surface of the filter and the negative zeta potential of the NPs. Such doped filters were used for the adsorption and reduction of 4-nitrophenol and methylene blue (MB), chosen as model water contaminants.

With a similar approach, Ismail et al. designed a naked eye sensor for Hg^{II}, Cr^{VI}, and ammonia in aqueous solution. They proceeded by dropping on a cellulose filter paper a colloidal solution of AgNPs, freshly prepared by using an aqueous leaf extract of *Convolvolus cneourm* to reduce Ag^+ ions and stabilize the resulting particles [179].

4.3. Cellulose Doping with AgNPs Generated in Situ

NPs can be also synthesized in situ and directly fixed on cellulose, which represents the preferred approach in many studies. As an example, a recent paper describes the synthesis of Ag_3PO_4 particles from $AgNO_3$ in the presence of NC sheets, following an ion exchange method [180]. The Ag_3PO_4/NC composite results in being particularly effective at promoting the sunlight-induced photodegradation of MB and methyl orange (MO), both in deionized water and in a more complex wastewater matrix.

More commonly, the immobilization of AgNPs on the cellulose support occurs by following the scheme depicted in Figure 7.

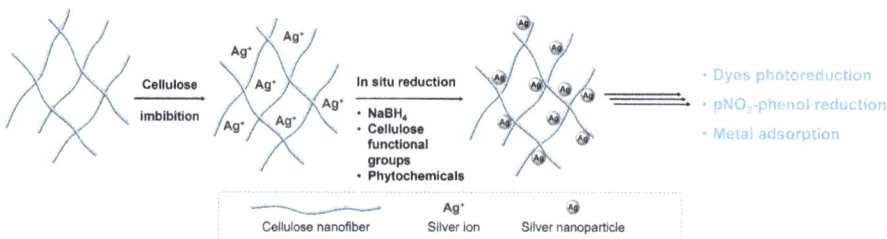

Figure 7. Process scheme for the in situ production of AgNPs/cellulose composites for water treatment.

The combination of different and complementary analytical techniques is necessary to confirm nanoparticle formation (XRD), bonding on the cellulose support by involving hydroxyl groups (FTIR), and homogeneous dispersion of nanoparticles on the polymeric matrix (SEM, TEM).

The standard procedure implies the impregnation of the polysaccharide substrate with an aqueous solution containing Ag^+ ions, usually provided in the form of $AgNO_3$ salt, and the consequent reduction of the metal ion with a proper reductant agent, like $NaBH_4$. Following this approach, Zia's group recently described an efficient adsorption process for the removal of metal ions from aqueous solutions, including wastewater, using AgNPs coated cotton cellulose as sorbent unit [181,182].

As the use of chemical reducing agents generates concerns, due to waste management issues, which in turn have a negative impact on the eco-design of final composites, in recent years the bioreduction solution has attracted more and more attention, being considered the safest and most convenient green option for AgNPs formation in situ. This consists of the use of phytochemicals like polyphenols, flavonoids, and alkaloids, extracted from different natural sources, as Ag$^+$ reductants. Albukari and coworkers have proposed the use of *Duranta erecta* leaves' extract for synthesizing AgNPs directly on cellulose polymer paper [20]. Das' group considered the *Hibiscus sabdariffa* shrub as a source of both nanocellulose from stem and reducing agents from flowers and leaves [183]. In both examples, the final composites were applied as effective photocatalysts under visible light radiation for the reduction of nitrophenols and a wide range of organic dyes in wastewater treatments. In a recent paper, Zhang and co-workers also proposed ascorbic acid as nontoxic reducing agent [184]. In this case, nanocomposites were synthesized through a double vegetable oil-based microemulsion method, using an ionic liquid (1-ethyl-3-methylimidazolium acetate) as a polar phase. Following this route, it was possible to simply control nanoparticle sizes and their homogeneous dispersion. The resulting material was successfully tested as catalyst for the reduction of nitrophenols in the presence of NaBH$_4$.

AgNP formation in situ could also be achieved without requiring the addition of external reductant agents. In 2013, Dong and coworkers reported the binding of Ag$^+$ ions onto TEMPO-oxidized CNF, followed by a slowly reduction of the cations by means of the hydroxyl groups of the glucopyranose units [185]. More recently, Pawcenis et al. deeply investigated this reduction mechanism [186]. Since the presence of aldehyde groups deriving by a partial oxidation has been excluded after both solution-state ^{1}H-NMR and solid-state ^{13}C-NMR, and FTIR analysis, they hypothesized that carbohydrates could play a similar role to diols, due to the high presence of vicinal hydroxy groups. Polyols can dehydrate to the corresponding unstable aldehydes, whose oxidation to carboxylic acids provides the electrons for the cation reduction.

Chook's group adopted this approach for the design of a porous AgNPs/CNF aerogel nanocomposite, which revealed superb performances in detecting and catalyzing the degradation of rhodamine B from water solutions, in the presence of NaBH$_4$ [112,187]. Similarly, Han et al. proposed the reductant-free synthesis of AgNPs-doped cellulose microgels for mediating the reduction of nitrophenols, always by means of NaBH$_4$ [188].

As stated before, one of the advantages in using NC supports consist into the significant increase of the reactive surface area, which also implies the possibility to functionalize further the cellulosic template in order to guarantee a stronger and more homogeneous binding of silver nanoparticles. A first approach could consist into simply adsorb functional moieties onto cellulose backbone. For example, An et al. showed how the adsorption of a surfactant (hexadecyl-tetramethylammonium bromide) onto CNC favors the binding and dispersion of AgNPs on the support, providing a device with a higher catalytic activity towards the reduction of 4-nitrophenol to 4-aminophenol, compared to AgNPs/CNC standard systems [189].

More commonly, functionalization occurs by covalently fixing new binding moieties. To design new membrane supports for water treatment, Nicolaus' group prepared amino-modified nanocellulose, using 3-aminopropyltriethoxysilane as a grafting agent, on which silver nanoparticles were fixed by in situ reduction with NaBH$_4$. The membrane performance was tested by measuring the total organic carbon (TOC) abatement from wastewater [190]. A greener approach to fix AgNPs on CNC was described by Tang and co-workers [191]. This consisted of the cellulose coating with mussel-inspired polydopamine (PDA), which also played the role of reducing agent for Ag$^+$ ions. The resulting AgNPs/PDA/CNC composite showed a core–shell structure decorated with nanoparticles. This system resulted in be six times more efficient in catalyzing the reduction of 4-nitrophenol, compared with the AgNPs/CNC model. The catalytic efficiency was even more accelerated by combing this CNC/PDA hybrid with β-cyclodextrin [191].

5. Conclusions and Perspectives

AgNPs' increasing integration into environmental applications, including water pollution monitoring and treatment, is raising concerns about their environmental impact, so much so that an eco-design strategy is proposed that combines better performance with the absence of risks for ecosystems. The grafting of AgNPs onto purposely selected polymers represents a potential solution for overcoming both ecosafety concerns and nanoparticle coalescence. In this context, with this contribution, starting from an overview of the recent preparations of AgNPs and their uses, we indicate in the (AgNPs)–cellulose hybrid materials potential solutions leading to efficient and eco-friendly materials for the monitoring and treatments of water pollution. (AgNPs)–cellulose hybrids have the double advantage of being easily produced using recycled material, with low cost and possible reuse, and of being eco-safe when properly designed. We envision that the development of these hybrid materials by following an eco-design approach would represent a winning strategy for valorizing AgNP-based nanotechnologies for water pollution monitoring and treatment.

Author Contributions: Conceptualization, original draft preparation, review and editing: A.F., C.P., I.C., A.B. and I.V.; supervision, I.V. All authors have read and agreed to the published version of the manuscript.

Funding: This research received no external funding.

Acknowledgments: The Grant of Excellence Departments, MIUR (ARTICOLO 1, COMMI 314–337 LEGGE 232/2016), is gratefully acknowledged by Iole Venditti.

Conflicts of Interest: The authors declare no conflict of interest.

References

1. Kumari, A.; Yadav, S.K. Nanotechnology in Agri-Food Sector. *Crit. Rev. Food Sci. Nutr.* **2014**, *54*, 975–984. [CrossRef] [PubMed]
2. Kim, J.; Van Der Bruggen, B. The use of nanoparticles in polymeric and ceramic membrane structures: Review of manufacturing procedures and performance improvement for water treatment. *Environ. Pollut.* **2010**, *158*, 2335–2349. [CrossRef] [PubMed]
3. Bearzotti, A.; Macagnano, A.; Papa, P.; Venditti, I.; Zampetti, E. A study of a QCM sensor based on pentacene for the detection of BTX vapors in air. *Sens. Actuators B Chem.* **2017**, *240*, 1160–1164. [CrossRef]
4. Yeshchenko, O.A.; Malynych, S.Z.; Polomarev, S.O.; Galabura, Y.; Chumanov, G.; Luzinov, I. Towards sensor applications of a polymer/Ag nanoparticle nanocomposite film. *RSC Adv.* **2019**, *9*, 8498–8506. [CrossRef]
5. Pantalei, S.; Zampetti, E.; Macagnano, A.; Bearzotti, A.; Venditti, I.; Russo, M.V. Enhanced Sensory Properties of a Multichannel Quartz Crystal Microbalance Coated with Polymeric Nanobeads. *Sensors* **2007**, *7*, 2920–2928. [CrossRef] [PubMed]
6. Yadav, A.; Kaushik, A.; Mishra, Y.K.; Agrawal, V.; Ahmadivand, A.; Maliutina, K.; Liu, Y.; Ouyang, Z.; Dong, W.; Cheng, G.J. Fabrication of 3D polymeric photonic arrays and related applications. *Mater. Today Chem.* **2020**, *15*. [CrossRef]
7. Bearzotti, A.; MacAgnano, A.; Pantalei, S.; Zampetti, E.; Venditti, I.; Fratoddi, I.; Vittoria Russo, M. Alcohol vapor sensory properties of nanostructured conjugated polymers. *J. Phys. Condens. Matter* **2008**, *20*. [CrossRef]
8. Rout, D.; Vijaya, R. Plasmonic Resonance-Induced Effects on Stopband and Emission Characteristics of Dye-Doped Opals. *Plasmonics* **2015**. [CrossRef]
9. De Angelis, R.; Venditti, I.; Fratoddi, I.; De Matteis, F.; Prosposito, P.; Cacciotti, I.; D'Amico, L.; Nanni, F.; Yadav, A.; Casalboni, M.; et al. From nanospheres to microribbons: Self-assembled Eosin Y doped PMMA nanoparticles as photonic crystals. *J. Colloid Interface Sci.* **2014**, *414*, 24–32. [CrossRef]
10. Atkuri, H.M.; Leong, E.S.P.; Hwang, J.; Palermo, G.; Si, G.; Wong, J.M.; Chien, L.C.; Ma, J.; Zhou, K.; Liu, Y.J.; et al. Developing novel liquid crystal technologies for display and photonic applications. *Displays* **2015**. [CrossRef]
11. Venditti, I.; D'Amato, R.; Russo, M.V.; Falconieri, M. Synthesis of conjugated polymeric nanobeads for photonic bandgap materials. *Sens. Actuators B Chem.* **2007**, *126*. [CrossRef]
12. Fratoddi, I. Hydrophobic and hydrophilic au and ag nanoparticles. Breakthroughs and perspectives. *Nanomaterials* **2018**, *8*, 11. [CrossRef] [PubMed]
13. Bonomo, M.; Naponiello, G.; Venditti, I.; Zardetto, V.; Di Carlo, A.; Dini, D. Electrochemical and Photoelectrochemical Properties of Screen-Printed Nickel Oxide Thin Films Obtained from Precursor Pastes with Different Compositions. *J. Electrochem. Soc.* **2017**, *164*, H137–H147. [CrossRef]

14. Busby, Y.; Agresti, A.; Pescetelli, S.; Di Carlo, A.; Noel, C.; Pireaux, J.-J.; Houssiau, L. Aging effects in interface-engineered perovskite solar cells with 2D nanomaterials: A depth profile analysis. *Mater. Today Energy* **2018**, *9*, 1–10. [CrossRef]
15. Naponiello, G.; Venditti, I.; Zardetto, V.; Saccone, D.; Di Carlo, A.; Fratoddi, I.; Barolo, C.; Dini, D. Photoelectrochemical characterization of squaraine-sensitized nickel oxide cathodes deposited via screen-printing for p-type dye-sensitized solar cells. *Appl. Surf. Sci.* **2015**, *356*. [CrossRef]
16. Caruso, M.; Gatto, E.; Palleschi, A.; Morales, P.; Scarselli, M.; Casaluci, S.; Quatela, A.; Di Carlo, A.; Venanzi, M. A bioinspired dye sensitized solar cell based on a rhodamine-functionalized peptide immobilized on nanocrystalline TiO_2. *J. Photochem. Photobiol. A Chem.* **2017**, *347*, 227–234. [CrossRef]
17. Vance, M.E.; Kuiken, T.; Vejerano, E.P.; McGinnis, S.P.; Hochella, M.F.; Hull, D.R. Nanotechnology in the real world: Redeveloping the nanomaterial consumer products inventory. *Beilstein J. Nanotechnol.* **2015**, *6*, 1769–1780. [CrossRef]
18. Navarro, E.; Wagner, B.; Odzak, N.; Sigg, L.; Behra, R. Effects of Differently Coated Silver Nanoparticles on the Photosynthesis of Chlamydomonas reinhardtii. *Environ. Sci. Technol.* **2015**, *49*, 8041–8047. [CrossRef]
19. Liu, B.; Wang, X.; Zhao, Y.; Wang, J.; Yang, X. Polymer shell as a protective layer for the sandwiched gold nanoparticles and their recyclable catalytic property. *J. Colloid Interface Sci.* **2013**. [CrossRef]
20. Albukhari, S.M.; Ismail, M.; Akhtar, K.; Danish, E.Y. Catalytic reduction of nitrophenols and dyes using silver nanoparticles @ cellulose polymer paper for the resolution of waste water treatment challenges. *Colloids Surf. A Physicochem. Eng. Asp.* **2019**, *577*, 548–561. [CrossRef]
21. Venditti, I. Morphologies and functionalities of polymeric nanocarriers as chemical tools for drug delivery: A review. *J. King Saud Univ. Sci.* **2019**, *31*, 398–411. [CrossRef]
22. Sánchez-López, E.; Gomes, D.; Esteruelas, G.; Bonilla, L.; Lopez-Machado, A.L.; Galindo, R.; Cano, A.; Espina, M.; Ettcheto, M.; Camins, A.; et al. Metal-based nanoparticles as antimicrobial agents: An overview. *Nanomaterials* **2020**, *10*, 292. [CrossRef] [PubMed]
23. Venditti, I.; Cartoni, A.; Fontana, L.; Testa, G.; Scaramuzzo, F.A.; Faccini, R.; Terracciano, C.M.; Camillocci, E.S.; Morganti, S.; Giordano, A.; et al. Y^{3+} embedded in polymeric nanoparticles: Morphology, dimension and stability of composite colloidal system. *Colloids Surf. A Physicochem. Eng. Asp.* **2017**, *532*. [CrossRef]
24. Hamidi-Asl, E.; Dardenne, F.; Pilehvar, S.; Blust, R.; De Wael, K. Unique properties of core shell Ag@Au nanoparticles for the aptasensing of bacterial cells. *Chemosensors* **2016**, *4*, 16. [CrossRef]
25. Lin, H.C.; Su, Y.A.; Liu, T.Y.; Sheng, Y.J.; Lin, J.J. Thermo-responsive nanoarrays of silver nanoparticle, silicate nanoplatelet and PNiPAAm for the antimicrobial applications. *Colloids Surf. B Biointerfaces* **2017**, *152*, 459–466. [CrossRef]
26. Mochi, F.; Burratti, L.; Fratoddi, I.; Venditti, I.; Battocchio, C.; Carlini, L.; Iucci, G.; Casalboni, M.; De Matteis, F.; Casciardi, S.; et al. Plasmonic sensor based on interaction between silver nanoparticles and Ni^{2+} or Co^{2+} in water. *Nanomaterials* **2018**, *8*, 488. [CrossRef]
27. Rajar, K.; Sirajuddin; Balouch, A.; Bhanger, M.I.; Shah, M.T.; Shaikh, T.; Siddiqui, S. Succinic acid functionalized silver nanoparticles (Suc-Ag NPs) for colorimetric sensing of melamine. *Appl. Surf. Sci.* **2018**, *435*, 1080–1086. [CrossRef]
28. Hjorth, R.; Coutris, C.; Nguyen, N.H.A.; Sevcu, A.; Gallego-Urrea, J.A.; Baun, A.; Joner, E.J. Ecotoxicity testing and environmental risk assessment of iron nanomaterials for sub-surface remediation—Recommendations from the FP7 project NanoRem. *Chemosphere* **2017**, *182*, 525–531. [CrossRef]
29. Corsi, I.; Cherr, G.N.; Lenihan, H.S.; Labille, J.; Hassellov, M.; Canesi, L.; Dondero, F.; Frenzilli, G.; Hristozov, D.; Puntes, V.; et al. Common strategies and technologies for the ecosafety assessment and design of nanomaterials entering the marine environment. *ACS Nano* **2014**, *8*, 9694–9709. [CrossRef]
30. Pham, T.L. Effect of Silver Nanoparticles on Tropical Freshwater and Marine Microalgae. *J. Chem.* **2019**, *2019*. [CrossRef]
31. Pulit-Prociak, J.; Banach, M. Silver nanoparticles—A material of the future...? *Open Chem.* **2016**, *14*, 76–91. [CrossRef]
32. Corsi, I.; Winther-Nielsen, M.; Sethi, R.; Punta, C.; Della Torre, C.; Libralato, G.; Lofrano, G.; Sabatini, L.; Aiello, M.; Fiordi, L.; et al. Ecofriendly nanotechnologies and nanomaterials for environmental applications: Key issue and consensus recommendations for sustainable and ecosafe nanoremediation. *Ecotoxicol. Environ. Saf.* **2018**, *154*, 237–244. [CrossRef] [PubMed]

33. MacCuspie, R.I. Colloidal stability of silver nanoparticles in biologically relevant conditions. *J. Nanopart. Res.* **2011**, *13*, 2893–2908. [CrossRef]
34. Xiang, Q.Q.; Wang, D.; Zhang, J.L.; Ding, C.Z.; Luo, X.; Tao, J.; Ling, J.; Shea, D.; Chen, L.Q. Effect of silver nanoparticles on gill membranes of common carp: Modification of fatty acid profile, lipid peroxidation and membrane fluidity. *Environ. Pollut.* **2020**, *256*. [CrossRef] [PubMed]
35. Prosposito, P.; Burratti, L.; Venditti, I. Silver Nanoparticles as Colorimetric Sensors for Water Pollutants. *Chemosensors* **2020**, *8*, 26. [CrossRef]
36. Marimuthu, S.; Antonisamy, A.J.; Malayandi, S.; Rajendran, K.; Tsai, P.C.; Pugazhendhi, A.; Ponnusamy, V.K. Silver nanoparticles in dye effluent treatment: A review on synthesis, treatment methods, mechanisms, photocatalytic degradation, toxic effects and mitigation of toxicity. *J. Photochem. Photobiol. B Biol.* **2020**, *205*, 111823. [CrossRef]
37. Kodoth, A.K.; Badalamoole, V. Silver nanoparticle-embedded pectin-based hydrogel for adsorptive removal of dyes and metal ions. *Polym. Bull.* **2020**, *77*, 541–564. [CrossRef]
38. Manimegalai, G.; Shanthakumar, S.; Sharma, C. Silver nanoparticles: synthesis and application in mineralization of pesticides using membrane support. *Int. Nano Lett.* **2014**, *4*. [CrossRef]
39. Kalantzi, I.; Mylona, K.; Toncelli, C.; Bucheli, T.D.; Knauer, K.; Pergantis, S.A.; Pitta, P.; Tsiola, A.; Tsapakis, M. Ecotoxicity of silver nanoparticles on plankton organisms: a review. *J. Nanopart. Res.* **2019**, *21*. [CrossRef]
40. Venditti, I. Engineered gold-based nanomaterials: Morphologies and functionalities in biomedical applications. a mini review. *Bioengineering* **2019**, *6*, 53. [CrossRef]
41. Bandala, E.R.; Berli, M. Engineered nanomaterials (ENMs) and their role at the nexus of Food, Energy, and Water. *Mater. Sci. Energy Technol.* **2019**, *2*, 29–40. [CrossRef]
42. Gubala, V.; Johnston, L.J.; Liu, Z.; Krug, H.; Moore, C.J.; Ober, C.K.; Schwenk, M.; Vert, M. Engineered nanomaterials and human health: Part 1. Preparation, functionalization and characterization (IUPAC Technical Report). *Pure Appl. Chem.* **2018**, *90*, 1283–1324. [CrossRef]
43. Janković, N.Z.; Plata, D.L. Engineered nanomaterials in the context of global element cycles. *Environ. Sci. Nano* **2019**, *6*, 2697–2711. [CrossRef]
44. Fratoddi, I.; Cartoni, A.; Venditti, I.; Catone, D.; O'Keeffe, P.; Paladini, A.; Toschi, F.; Turchini, S.; Sciubba, F.; Testa, G.; et al. Gold nanoparticles functionalized by rhodamine B isothiocyanate: A new tool to control plasmonic effects. *J. Colloid Interface Sci.* **2018**, *513*, 10–19. [CrossRef] [PubMed]
45. Wang, X.; Wu, J.; Li, P.; Wang, L.; Zhou, J.; Zhang, G.; Li, X.; Hu, B.; Xing, X. Microenvironment-Responsive Magnetic Nanocomposites Based on Silver Nanoparticles/Gentamicin for Enhanced Biofilm Disruption by Magnetic Field. *ACS Appl. Mater. Interfaces* **2018**, *10*, 34905–34915. [CrossRef] [PubMed]
46. Catone, D.; Ciavardini, A.; Di Mario, L.; Paladini, A.; Toschi, F.; Cartoni, A.; Fratoddi, I.; Venditti, I.; Alabastri, A.; Proietti Zaccaria, R.; et al. Plasmon Controlled Shaping of Metal Nanoparticle Aggregates by Femtosecond Laser-Induced Melting. *J. Phys. Chem. Lett.* **2018**, *9*, 5002–5008. [CrossRef] [PubMed]
47. Bracalello, A.; Secchi, V.; Mastrantonio, R.; Pepe, A.; Persichini, T.; Iucci, G.; Bochicchio, B.; Battocchio, C. Fibrillar self-assembly of a chimeric elastin-resilin inspired engineered polypeptide. *Nanomaterials* **2019**, *9*, 1613. [CrossRef]
48. Schiesaro, I.; Battocchio, C.; Venditti, I.; Prosposito, P.; Burratti, L.; Centomo, P.; Meneghini, C. Structural characterization of 3d metal adsorbed AgNPs. *Phys. E Low-Dimens. Syst. Nanostructures* **2020**, *123*, 114162. [CrossRef]
49. Venditti, I.; Testa, G.; Sciubba, F.; Carlini, L.; Porcaro, F.; Meneghini, C.; Mobilio, S.; Battocchio, C.; Fratoddi, I. Hydrophilic Metal Nanoparticles Functionalized by 2-Diethylaminoethanethiol: A Close Look at the Metal-Ligand Interaction and Interface Chemical Structure. *J. Phys. Chem. C* **2017**, *121*. [CrossRef]
50. Kailasa, S.K.; Koduru, J.R.; Desai, M.L.; Park, T.J.; Singhal, R.K.; Basu, H. Recent progress on surface chemistry of plasmonic metal nanoparticles for colorimetric assay of drugs in pharmaceutical and biological samples. *TrAC Trends Anal. Chem.* **2018**, *105*, 106–120. [CrossRef]
51. Yoon, S.J.; Nam, Y.S.; Lee, H.J.; Lee, Y.; Lee, K.B. Colorimetric probe for Ni^{2+} based on shape transformation of triangular silver nanoprisms upon H_2O_2 etching. *Sens. Actuators B Chem.* **2019**, *300*. [CrossRef]
52. Farhadi, K.; Forough, M.; Molaei, R.; Hajizadeh, S.; Rafipour, A. Highly selective Hg^{2+} colorimetric sensor using green synthesized and unmodified silver nanoparticles. *Sens. Actuators B Chem.* **2012**, *161*, 880–885. [CrossRef]

53. Loganathan, C.; John, S.A. Naked eye and spectrophotometric detection of chromogenic insecticide in aquaculture using amine functionalized gold nanoparticles in the presence of major interferents. *Spectrochim. Acta Part A Mol. Biomol. Spectrosc.* **2017**, *173*, 837–842. [CrossRef] [PubMed]
54. Prosposito, P.; Burratti, L.; Bellingeri, A.; Protano, G.; Faleri, C.; Corsi, I.; Battocchio, C.; Iucci, G.; Tortora, L.; Secchi, V.; et al. Bifunctionalized silver nanoparticles as Hg^{2+} plasmonic sensor in water: Synthesis, characterizations, and ecosafety. *Nanomaterials* **2019**, *9*, 1353. [CrossRef]
55. Rahim, S.; Khalid, S.; Bhanger, M.I.; Shah, M.R.; Malik, M.I. Polystyrene-block-poly(2-vinylpyridine) -conjugated silver nanoparticles as colorimetric sensor for quantitative determination of Cartap in aqueous media and blood plasma. *Sens. Actuators B Chem.* **2018**, *259*, 878–887. [CrossRef]
56. Vilela, D.; González, M.C.; Escarpa, A. Sensing colorimetric approaches based on gold and silver nanoparticles aggregation: Chemical creativity behind the assay. A review. *Anal. Chim. Acta* **2012**, *751*, 24–43. [CrossRef]
57. Rycenga, M.; Cobley, C.M.; Zeng, J.; Li, W.; Moran, C.H.; Zhang, Q.; Qin, D.; Xia, Y. Controlling the synthesis and assembly of silver nanostructures for plasmonic applications. *Chem. Rev.* **2011**, *111*, 3669–3712. [CrossRef]
58. Wiley, B.; Sun, Y.; Xia, Y. Synthesis of silver nanostructures with controlled shapes and properties. *Acc. Chem. Res.* **2007**, *40*, 1067–1076. [CrossRef]
59. Link, S.; El-Sayed, M.A. Spectral Properties and Relaxation Dynamics of Surface Plasmon Electronic Oscillations in Gold and Silver Nanodots and Nanorods. *J. Phys. Chem. B* **1999**, *103*, 8410–8426. [CrossRef]
60. Cao, W.; Elsayed-Ali, H.E. Stability of Ag nanoparticles fabricated by electron beam lithography. *Mater. Lett.* **2009**, *63*, 2263–2266. [CrossRef]
61. Melinte, V.; Stroea, L.; Buruiana, T.; Chibac, A.L. Photocrosslinked hybrid composites with Ag, Au or Au-Ag NPs as visible-light triggered photocatalysts for degradation/reduction of aromatic nitroderivatives. *Eur. Polym. J.* **2019**, *121*. [CrossRef]
62. Emam, H.E.; El-Zawahry, M.M.; Ahmed, H.B. One-pot fabrication of AgNPs, AuNPs and Ag-Au nano-alloy using cellulosic solid support for catalytic reduction application. *Carbohydr. Polym.* **2017**, *166*, 1–13. [CrossRef] [PubMed]
63. Farooqi, Z.H.; Khalid, R.; Begum, R.; Farooq, U.; Wu, Q.; Wu, W.; Ajmal, M.; Irfan, A.; Naseem, K. Facile synthesis of silver nanoparticles in a crosslinked polymeric system by in situ reduction method for catalytic reduction of 4-nitroaniline. *Environ. Technol. (UK)* **2019**, *40*, 2027–2036. [CrossRef] [PubMed]
64. Battocchio, C.; Meneghini, C.; Fratoddi, I.; Venditti, I.; Russo, M.V.; Aquilanti, G.; Maurizio, C.; Bondino, F.; Matassa, R.; Rossi, M.; et al. Silver nanoparticles stabilized with thiols: A close look at the local chemistry and chemical structure. *J. Phys. Chem. C* **2012**, *116*. [CrossRef]
65. Hsu, C.W.; Lin, Z.Y.; Chan, T.Y.; Chiu, T.C.; Hu, C.C. Oxidized multiwalled carbon nanotubes decorated with silver nanoparticles for fluorometric detection of dimethoate. *Food Chem.* **2017**, *224*, 353–358. [CrossRef]
66. Slepička, P.; Kasálková, N.S.; Siegel, J.; Kolská, Z.; Švorčík, V. Methods of gold and silver nanoparticles preparation. *Materials (Basel)* **2020**, *13*, 1.
67. Basiri, S.; Mehdinia, A.; Jabbari, A. Biologically green synthesized silver nanoparticles as a facile and rapid label-free colorimetric probe for determination of Cu2 + in water samples. *Spectrochim. Acta Part A Mol. Biomol. Spectrosc.* **2017**, *171*, 297–304. [CrossRef]
68. Corsi, P.; Venditti, I.; Battocchio, C.; Meneghini, C.; Bruni, F.; Prosposito, P.; Mochi, F.; Capone, B. Designing an optimal ion adsorber on the nanoscale: a simple theoretical model for the unusual nucleation of AgNPs/Co^{2+}–Ni^{2+} binary mixtures. *J. Phys. Chem. C* **2019**, *123*, 3855–3860; [CrossRef]
69. Rauwel, P.; Rauwel, E.; Ferdov, S.; Singh, M.P. Silver nanoparticles: Synthesis, properties, and applications. *Adv. Mater. Sci. Eng.* **2015**, *2015*. [CrossRef]
70. Liu, J.; Sonshine, D.A.; Shervani, S.; Hurt, R.H. Controlled release of biologically active silver from nanosilver surfaces. *ACS Nano* **2010**, *4*, 6903–6913. [CrossRef]
71. Carlini, L.; Fasolato, C.; Postorino, P.; Fratoddi, I.; Venditti, I.; Testa, G.; Battocchio, C. Comparison between silver and gold nanoparticles stabilized with negatively charged hydrophilic thiols: SR-XPS and SERS as probes for structural differences and similarities. *Colloids Surfaces A Physicochem. Eng. Asp.* **2017**, *532*. [CrossRef]
72. Dong, Y.; Ding, L.; Jin, X.; Zhu, N. Silver nanoparticles capped with chalcon carboxylic acid as a probe for colorimetric determination of cadmium(II). *Microchim. Acta* **2017**, *184*, 3357–3362. [CrossRef]
73. Roto, R.; Mellisani, B.; Kuncaka, A.; Mudasir, M.; Suratman, A. Colorimetric sensing of Pb^{2+} ion by using ag nanoparticles in the presence of dithizone. *Chemosensors* **2019**, *7*, 28. [CrossRef]

74. Ratnarathorn, N.; Chailapakul, O.; Henry, C.S.; Dungchai, W. Simple silver nanoparticle colorimetric sensing for copper by paper-based devices. *Talanta* **2012**, *99*, 552–557. [CrossRef] [PubMed]
75. Roy, K.; Sarkar, C.K.; Ghosh, C.K. Rapid colorimetric detection of Hg^{2+} ion by green silver nanoparticles synthesized using Dahlia pinnata leaf extract. *Green Process. Synth.* **2015**, *4*, 455–461. [CrossRef]
76. Choudhury, R.; Misra, T.K. Gluconate stabilized silver nanoparticles as a colorimetric sensor for Pb^{2+}. *Colloids Surf. A Physicochem. Eng. Asp.* **2018**, *545*, 179–187. [CrossRef]
77. Zheng, M.; He, J.; Wang, Y.; Wang, C.; Ma, S.; Sun, X. Colorimetric recognition of 6-benzylaminopurine in environmental samples by using thioglycolic acid functionalized silver nanoparticles. *Spectrochim. Acta Part A Mol. Biomol. Spectrosc.* **2018**, *192*, 27–33. [CrossRef]
78. Bala, R.; Mittal, S.; Sharma, R.K.; Wangoo, N. A supersensitive silver nanoprobe based aptasensor for low cost detection of malathion residues in water and food samples. *Spectrochim. Acta Part A Mol. Biomol. Spectrosc.* **2018**, *196*, 268–273. [CrossRef]
79. Roy, B.; Bairi, P.; Nandi, A.K. Selective colorimetric sensing of mercury(ii) using turn off-turn on mechanism from riboflavin stabilized silver nanoparticles in aqueous medium. *Analyst* **2011**, *136*, 3605–3607. [CrossRef]
80. Wang, B.; Zhang, L.; Zhou, X. Synthesis of silver nanocubes as a SERS substrate for the determination of pesticide paraoxon and thiram. *Spectrochim. Acta Part A Mol. Biomol. Spectrosc.* **2014**, *121*, 63–69. [CrossRef]
81. Han, H.J.; Yu, T.; Kim, W.S.; Im, S.H. Highly reproducible polyol synthesis for silver nanocubes. *J. Cryst. Growth* **2017**, *469*, 48–53. [CrossRef]
82. Zhou, S.; Li, J.; Gilroy, K.D.; Tao, J.; Zhu, C.; Yang, X.; Sun, X.; Xia, Y. Facile Synthesis of Silver Nanocubes with Sharp Corners and Edges in an Aqueous Solution. *ACS Nano* **2016**, *10*, 9861–9870. [CrossRef] [PubMed]
83. Liu, L.; Wu, Y.; Yin, N.; Zhang, H.; Ma, H. Silver nanocubes with high SERS performance. *J. Quant. Spectrosc. Radiat. Transf.* **2020**, *240*. [CrossRef]
84. Swaminathan, N.; Nerthigan, Y.; Wu, H.-F. Polyaniline stabilized Silver (I) Oxide nanocubes for sensitive and selective detection of hemoglobin in urine for hematuria evaluation. *Microchem. J.* **2020**, *155*. [CrossRef]
85. Chakraborty, U.; Bhanjana, G.; Kaur, G.; Kaushik, A.; Chaudhary, G.R. Electro-active silver oxide nanocubes for label free direct sensing of bisphenol A to assure water quality. *Mater. Today Chem.* **2020**, *16*. [CrossRef]
86. Joseph, D.; Huh, Y.S.; Han, Y.K. A top-down chemical approach to tuning the morphology and plasmon resonance of spiky nanostars for enriched SERS-based chemical sensing. *Sens. Actuators B Chem.* **2019**, *288*, 120–126. [CrossRef]
87. Mulder, D.W.; Phiri, M.M.; Vorster, B.C. Tailor-made gold nanostar colorimetric detection determined by morphology change and used as an indirect approach by using hydrogen peroxide to determine glucose concentration. *Sens. Bio-Sens. Res.* **2019**, *25*, 100296. [CrossRef]
88. Garcia-Leis, A.; Rivera-Arreba, I.; Sanchez-Cortes, S. Morphological tuning of plasmonic silver nanostars by controlling the nanoparticle growth mechanism: Application in the SERS detection of the amyloid marker Congo Red. *Colloids Surfaces A Physicochem. Eng. Asp.* **2017**, *535*, 49–60. [CrossRef]
89. de Almeida, M.P.; Leopold, N.; Franco, R.; Pereira, E. Expedite SERS fingerprinting of Portuguese white wines using plasmonic silver nanostars. *Front. Chem.* **2019**, *7*, 1–9. [CrossRef]
90. Zhang, W.; Liu, J.; Niu, W.; Yan, H.; Lu, X.; Liu, B. Tip-Selective Growth of Silver on Gold Nanostars for Surface-Enhanced Raman Scattering. *ACS Appl. Mater. Interfaces* **2018**, *10*, 14850–14856. [CrossRef]
91. Joseph, D.; Baskaran, R.; Yang, S.G.; Huh, Y.S.; Han, Y.K. Multifunctional spiky branched gold-silver nanostars with near-infrared and short-wavelength infrared localized surface plasmon resonances. *J. Colloid Interface Sci.* **2019**, *542*, 308–316. [CrossRef] [PubMed]
92. Zhu, J.; Chen, X.H.; Li, J.J.; Zhao, J.W. The synthesis of Ag-coated tetrapod gold nanostars and the improvement of surface-enhanced Raman scattering. *Spectrochim. Acta Part A Mol. Biomol. Spectrosc.* **2019**, *211*, 154–165. [CrossRef] [PubMed]
93. Reyes Gómez, F.; Rubira, R.J.G.; Camacho, S.A.; Martin, C.S.; da Silva, R.R.; Constantino, C.J.L.; Alessio, P.; Oliveira, O.N.; Mejía-Salazar, J.R. Surface Plasmon Resonances in Silver Nanostars. *Sensors (Basel)* **2018**, *18*, 3821. [CrossRef]
94. Lee, S.H.; Jun, B.H. Silver nanoparticles: Synthesis and application for nanomedicine. *Int. J. Mol. Sci.* **2019**, *20*, 865. [CrossRef]
95. Jakab, A.; Rosman, C.; Khalavka, Y.; Becker, J.; Trügler, A.; Hohenester, U.; Sönnichsen, C. Highly sensitive plasmonic silver nanorods. *ACS Nano* **2011**, *5*, 6880–6885. [CrossRef]

96. Maccora, D.; Dini, V.; Battocchio, C.; Fratoddi, I.; Cartoni, A.; Rotili, D.; Castagnola, M.; Faccini, R.; Bruno, I.; Scotognella, T.; et al. Gold nanoparticles and nanorods in nuclear medicine: A mini review. *Appl. Sci.* **2019**, *9*, 3232. [CrossRef]
97. Zhan, S.; Yu, M.; Lv, J.; Wang, L.; Zhou, P. Colorimetric detection of trace arsenic(III) in aqueous solution using arsenic aptamer and gold nanoparticles. *Aust. J. Chem.* **2014**, *67*, 813–818. [CrossRef]
98. Rekha, C.R.; Nayar, V.U.; Gopchandran, K.G. Synthesis of highly stable silver nanorods and their application as SERS substrates. *J. Sci. Adv. Mater. Devices* **2018**, *3*, 196–205. [CrossRef]
99. Liu, Y.J.; Chu, H.Y.; Zhao, Y.P. Silver nanorod array substrates fabricated by oblique angle deposition: Morphological, optical, and SERS characterizations. *J. Phys. Chem. C* **2010**, *114*, 8176–8183. [CrossRef]
100. He, Y.; Fu, J.; Zhao, Y. Oblique angle deposition and its applications in plasmonics. *Front. Phys.* **2014**, *9*, 47–59. [CrossRef]
101. Qu, F.; Chen, P.; Zhu, S.; You, J. High selectivity of colorimetric detection of p-nitrophenol based on Ag nanoclusters. *Spectrochim. Acta Part A Mol. Biomol. Spectrosc.* **2017**, *171*, 449–453. [CrossRef] [PubMed]
102. Rohit, J.V.; Kailasa, S.K. Cyclen dithiocarbamate-functionalized silver nanoparticles as a probe for colorimetric sensing of thiram and paraquat pesticides via host–guest chemistry. *J. Nanopart. Res.* **2014**, *16*. [CrossRef]
103. Qu, L.L.; Geng, Z.Q.; Wang, W.; Yang, K.C.; Wang, W.P.; Han, C.Q.; Yang, G.H.; Vajtai, R.; Li, D.W.; Ajayan, P.M. Recyclable three-dimensional Ag nanorod arrays decorated with O-g-C_3N_4 for highly sensitive SERS sensing of organic pollutants. *J. Hazard. Mater.* **2019**, *379*. [CrossRef] [PubMed]
104. Song, C.; Yang, B.; Zhu, Y.; Yang, Y.; Wang, L. Ultrasensitive sliver nanorods array SERS sensor for mercury ions. *Biosens. Bioelectron.* **2017**, *87*, 59–65. [CrossRef]
105. Zhou, Q.; Yang, Y.; Ni, J.; Li, Z.; Zhang, Z. Rapid detection of 2, 3, 3′, 4, 4′-pentachlorinated biphenyls by silver nanorods-enhanced Raman spectroscopy. *Phys. E Low-Dimensional Syst. Nanostructures* **2010**, *42*, 1717–1720. [CrossRef]
106. Shkilnyy, A.; Soucé, M.; Dubois, P.; Warmont, F.; Saboungi, M.-L.; Chourpa, I. Poly(ethylene glycol)-stabilized silver nanoparticles for bioanalytical applications of SERS spectroscopy. *Analyst* **2009**, *134*, 1868–1872. [CrossRef]
107. Manivannan, S.; Ramaraj, R. Silver nanoparticles embedded in cyclodextrin-silicate composite and their applications in Hg(ii) ion and nitrobenzene sensing. *Analyst* **2013**, *138*, 1733–1739. [CrossRef]
108. Sharma, P.; Mourya, M.; Choudhary, D.; Goswami, M.; Kundu, I.; Dobhal, M.P.; Tripathi, C.S.P.; Guin, D. Thiol terminated chitosan capped silver nanoparticles for sensitive and selective detection of mercury (II) ions in water. *Sens. Actuators B Chem.* **2018**, *268*, 310–318. [CrossRef]
109. Ullah, S.; Fayeza; Khan, A.A.; Jan, A.; Aain, S.Q.; Neto, E.P.F.; Serge-Correales, Y.E.; Parveen, R.; Wender, H.; Rodrigues-Filho, U.P.; et al. Enhanced photoactivity of BiVO4/Ag/Ag2O Z-scheme photocatalyst for efficient environmental remediation under natural sunlight and low-cost LED illumination. *Colloids Surf. A Physicochem. Eng. Asp.* **2020**, *600*. [CrossRef]
110. Alamelu, K.; Jaffar Ali, B.M. Ag nanoparticle-impregnated sulfonated graphene/TiO_2 composite for the photocatalytic removal of organic pollutants. *Appl. Surf. Sci.* **2020**, *512*. [CrossRef]
111. Kodom, T.; Rusen, E.; Călinescu, I.; Mocanu, A.; Şomoghi, R.; Dinescu, A.; Diacon, A.; Boscornea, C. Silver Nanoparticles Influence on Photocatalytic Activity of Hybrid Materials Based on TiO_2 P25. *J. Nanomater.* **2015**, *2015*. [CrossRef]
112. Chook, S.W.; Chia, C.H.; Chan, C.H.; Chin, S.X.; Zakaria, S.; Sajab, M.S.; Huang, N.M. A porous aerogel nanocomposite of silver nanoparticles-functionalized cellulose nanofibrils for SERS detection and catalytic degradation of rhodamine B. *RSC Adv.* **2015**, *5*, 88915–88920. [CrossRef]
113. Porcaro, F.; Carlini, L.; Ugolini, A.; Visaggio, D.; Visca, P.; Fratoddi, I.; Venditti, I.; Meneghini, C.; Simonelli, L.; Marini, C.; et al. Synthesis and structural characterization of silver nanoparticles stabilized with 3-mercapto-1-propansulfonate and 1-thioglucose mixed thiols for antibacterial applications. *Materials (Basel)* **2016**, *9*, 28. [CrossRef] [PubMed]
114. Blasco, J.; Corsi, I. *Ecotoxicology of Nanoparticles in Aquatic Systems*; CRC Press: Boca Raton, FL, USA, 2019.
115. Nel, A.; Xia, T.; Mädler, L.; Li, N. Toxic potential of materials at the nanolevel. *Science* **2006**, *311*, 622–627. [CrossRef]
116. Albanese, A.; Tang, P.S.; Chan, W.C.W. The Effect of Nanoparticle Size, Shape, and Surface Chemistry on Biological Systems. *Annu. Rev. Biomed. Eng.* **2012**, *14*, 1–16. [CrossRef] [PubMed]

117. Baalousha, M.; Cornelis, G.; Kuhlbusch, T.A.J.; Lynch, I.; Nickel, C.; Peijnenburg, W.; Van Den Brink, N.W. Modeling nanomaterial fate and uptake in the environment: Current knowledge and future trends. *Environ. Sci. Nano* **2016**, *3*, 323–345. [CrossRef]
118. Garner, K.L.; Suh, S.; Keller, A.A. Assessing the Risk of Engineered Nanomaterials in the Environment: Development and Application of the nanoFate Model. *Environ. Sci. Technol.* **2017**, *51*, 5541–5551. [CrossRef]
119. Praetorius, A.; Badetti, E.; Brunelli, A.; Clavier, A.; Gallego-Urrea, J.A.; Gondikas, A.; Hassellöv, M.; Hofmann, T.; Mackevica, A.; Marcomini, A.; et al. Strategies for determining heteroaggregation attachment efficiencies of engineered nanoparticles in aquatic environments. *Environ. Sci. Nano* **2020**, *7*, 351–367. [CrossRef]
120. Therezien, M.; Thill, A.; Wiesner, M.R. Importance of heterogeneous aggregation for NP fate in natural and engineered systems. *Sci. Total Environ.* **2014**, *485–486*, 309–318. [CrossRef]
121. Mićić, V.; Bossa, N.; Schmid, D.; Wiesner, M.R.; Hofmann, T. Groundwater Chemistry Has a Greater Influence on the Mobility of Nanoparticles Used for Remediation than the Chemical Heterogeneity of Aquifer Media. *Environ. Sci. Technol.* **2020**, *54*, 1250–1257. [CrossRef]
122. Jorge de Souza, T.A.; Rosa Souza, L.R.; Franchi, L.P. Silver nanoparticles: An integrated view of green synthesis methods, transformation in the environment, and toxicity. *Ecotoxicol. Environ. Saf.* **2019**, *171*, 691–700. [CrossRef] [PubMed]
123. Silver, S.; Phung, L.T.; Silver, G. Silver as biocides in burn and wound dressings and bacterial resistance to silver compounds. *J. Ind. Microbiol. Biotechnol.* **2006**, *33*, 627–634. [CrossRef] [PubMed]
124. Kaiser, J.P.; Zuin, S.; Wick, P. Is nanotechnology revolutionizing the paint and lacquer industry? A critical opinion. *Sci. Total Environ.* **2013**, *442*, 282–289. [CrossRef] [PubMed]
125. Dastjerdi, R.; Montazer, M. A review on the application of inorganic nano-structured materials in the modification of textiles: Focus on anti-microbial properties. *Colloids Surf. B Biointerfaces* **2010**, *79*, 5–18. [CrossRef]
126. Li, L.; Stoiber, M.; Wimmer, A.; Xu, Z.; Lindenblatt, C.; Helmreich, B.; Schuster, M. To What Extent Can Full-Scale Wastewater Treatment Plant Effluent Influence the Occurrence of Silver-Based Nanoparticles in Surface Waters? *Environ. Sci. Technol.* **2016**, *50*, 6327–6333. [CrossRef]
127. McGillicuddy, E.; Murray, I.; Kavanagh, S.; Morrison, L.; Fogarty, A.; Cormican, M.; Dockery, P.; Prendergast, M.; Rowan, N.; Morris, D. Silver nanoparticles in the environment: Sources, detection and ecotoxicology. *Sci. Total Environ.* **2017**, *575*, 231–246. [CrossRef]
128. Tortella, G.R.; Rubilar, O.; Durán, N.; Diez, M.C.; Martínez, M.; Parada, J.; Seabra, A.B. Silver nanoparticles: Toxicity in model organisms as an overview of its hazard for human health and the environment. *J. Hazard. Mater.* **2020**, *390*, 121974. [CrossRef]
129. Liu, H.; Wang, X.; Wu, Y.; Hou, J.; Zhang, S.; Zhou, N.; Wang, X. Toxicity responses of different organs of zebrafish (Danio rerio) to silver nanoparticles with different particle sizes and surface coatings. *Environ. Pollut.* **2019**, *246*, 414–422. [CrossRef]
130. Kleiven, M.; Macken, A.; Oughton, D.H. Growth inhibition in Raphidocelis subcapita—Evidence of nanospecific toxicity of silver nanoparticles. *Chemosphere* **2019**, *221*, 785–792. [CrossRef]
131. Pang, C.; Brunelli, A.; Zhu, C.; Hristozov, D.; Liu, Y.; Semenzin, E.; Wang, W.; Tao, W.; Liang, J.; Marcomini, A.; et al. Demonstrating approaches to chemically modify the surface of Ag nanoparticles in order to influence their cytotoxicity and biodistribution after single dose acute intravenous administration. *Nanotoxicology* **2016**, *10*, 129–139. [CrossRef]
132. Dong, F.; Mohd Zaidi, N.F.; Valsami-Jones, E.; Kreft, J.U. Time-resolved toxicity study reveals the dynamic interactions between uncoated silver nanoparticles and bacteria. *Nanotoxicology* **2017**, *11*, 637–646. [CrossRef] [PubMed]
133. Ratte, H.T. Bioaccumulation and toxicity of silver compounds: A review. *Environ. Toxicol. Chem.* **1999**, *18*, 89–108. [CrossRef]
134. Zhang, C.; Hu, Z.; Deng, B. Silver nanoparticles in aquatic environments: Physiochemical behavior and antimicrobial mechanisms. *Water Res.* **2016**, *88*, 403–427. [CrossRef] [PubMed]
135. Rai, M.; Yadav, A.; Gade, A. Silver nanoparticles as a new generation of antimicrobials. *Biotechnol. Adv.* **2009**, *27*, 76–83. [CrossRef] [PubMed]

136. Lekamge, S.; Miranda, A.F.; Pham, B.; Ball, A.S.; Shukla, R.; Nugegoda, D. The toxicity of non-aged and aged coated silver nanoparticles to the freshwater shrimp Paratya australiensis. *J. Toxicol. Environ. Health Part A Curr. Issues* **2019**, *82*, 1207–1222. [CrossRef]
137. Lekamge, S.; Miranda, A.F.; Trestrail, C.; Pham, B.; Ball, A.S.; Shukla, R.; Nugegoda, D. The Toxicity of Nonaged and Aged Coated Silver Nanoparticles to Freshwater Alga Raphidocelis subcapitata. *Environ. Toxicol. Chem.* **2019**, *38*, 2371–2382. [CrossRef]
138. Sendra, M.; Yeste, M.P.; Gatica, J.M.; Moreno-Garrido, I.; Blasco, J. Direct and indirect effects of silver nanoparticles on freshwater and marine microalgae (Chlamydomonas reinhardtii and Phaeodactylum tricornutum). *Chemosphere* **2017**, *179*, 279–289. [CrossRef]
139. Carrazco-Quevedo, A.; Römer, I.; Salamanca, M.J.; Poynter, A.; Lynch, I.; Valsami-Jones, E. Bioaccumulation and toxic effects of nanoparticulate and ionic silver in Saccostrea glomerata (rock oyster). *Ecotoxicol. Environ. Saf.* **2019**, *179*, 127–134. [CrossRef]
140. An, H.J.; Sarkheil, M.; Park, H.S.; Yu, I.J.; Johari, S.A. Comparative toxicity of silver nanoparticles (AgNPs) and silver nanowires (AgNWs) on saltwater microcrustacean, Artemia salina. *Comp. Biochem. Physiol. Part C Toxicol. Pharmacol.* **2019**, *218*, 62–69. [CrossRef]
141. Ale, A.; Liberatori, G.; Vannuccini, M.L.; Bergami, E.; Ancora, S.; Mariotti, G.; Bianchi, N.; Galdopórpora, J.M.; Desimone, M.F.; Cazenave, J.; et al. Exposure to a nanosilver-enabled consumer product results in similar accumulation and toxicity of silver nanoparticles in the marine mussel Mytilus galloprovincialis. *Aquat. Toxicol.* **2019**, *211*, 46–56. [CrossRef]
142. Gliga, A.R.; Skoglund, S.; Odnevall Wallinder, I.; Fadeel, B.; Karlsson, H.L. Size-dependent cytotoxicity of silver nanoparticles in human lung cells: The role of cellular uptake, agglomeration and Ag release. *Part. Fibre Toxicol.* **2014**, *11*. [CrossRef] [PubMed]
143. Ivask, A.; Kurvet, I.; Kasemets, K.; Blinova, I.; Aruoja, V.; Suppi, S.; Vija, H.; Käkinen, A.; Titma, T.; Heinlaan, M.; et al. Size-dependent toxicity of silver nanoparticles to bacteria, yeast, algae, crustaceans and mammalian cells in vitro. *PLoS ONE* **2014**, *9*. [CrossRef] [PubMed]
144. Guo, Z.; Zeng, G.; Cui, K.; Chen, A. Toxicity of environmental nanosilver: mechanism and assessment. *Environ. Chem. Lett.* **2019**, *17*, 319–333. [CrossRef]
145. Radwan, I.M.; Gitipour, A.; Potter, P.M.; Dionysiou, D.D.; Al-Abed, S.R. Dissolution of silver nanoparticles in colloidal consumer products: Effects of particle size and capping agent. *J. Nanopart. Res.* **2019**, *21*. [CrossRef] [PubMed]
146. Pem, B.; Pongrac, I.M.; Ulm, L.; Pavičić, I.; Vrček, V.; Jurašin, D.D.; Ljubojević, M.; Krivohlavek, A.; Vrček, I.V. Toxicity and safety study of silver and gold nanoparticles functionalized with cysteine and glutathione. *Beilstein J. Nanotechnol.* **2019**, *10*, 1802–1817. [CrossRef]
147. Yu, S.J.; Lai, Y.J.; Dong, L.J.; Liu, J.F. Intracellular Dissolution of Silver Nanoparticles: Evidence from Double Stable Isotope Tracing. *Environ. Sci. Technol.* **2019**, *53*, 10218–10226. [CrossRef]
148. Angel, B.M.; Batley, G.E.; Jarolimek, C.V.; Rogers, N.J. The impact of size on the fate and toxicity of nanoparticulate silver in aquatic systems. *Chemosphere* **2013**, *93*, 359–365. [CrossRef]
149. Auclair, J.; Turcotte, P.; Gagnon, C.; Peyrot, C.; Wilkinson, K.J.; Gagné, F. The influence of surface coatings on the toxicity of silver nanoparticle in rainbow trout. *Comp. Biochem. Physiol. Part C Toxicol. Pharmacol.* **2019**, *226*, 108623. [CrossRef]
150. Luoma, S.N. Silver nanotechnologies and the environment: Old problems or new challenges? *Proj. Emerg. Nanotechnol. Rep.* **2008**, *15*, 12–13.
151. Gunsolus, I.L.; Mousavi, M.P.S.; Hussein, K.; Bühlmann, P.; Haynes, C.L. Effects of Humic and Fulvic Acids on Silver Nanoparticle Stability, Dissolution, and Toxicity. *Environ. Sci. Technol.* **2015**, *49*, 8078–8086. [CrossRef]
152. Ding, Y.; Bai, X.; Ye, Z.; Gong, D.; Cao, J.; Hua, Z. Humic acid regulation of the environmental behavior and phytotoxicity of silver nanoparticles to: Lemna minor. *Environ. Sci. Nano* **2019**, *6*, 3712–3722. [CrossRef]
153. Cáceres-Vélez, P.R.; Fascineli, M.L.; Sousa, M.H.; Grisolia, C.K.; Yate, L.; de Souza, P.E.N.; Estrela-Lopis, I.; Moya, S.; Azevedo, R.B. Humic acid attenuation of silver nanoparticle toxicity by ion complexation and the formation of a Ag^{3+} coating. *J. Hazard. Mater.* **2018**, *353*, 173–181. [CrossRef] [PubMed]
154. Levard, C.; Hotze, E.M.; Colman, B.P.; Dale, A.L.; Truong, L.; Yang, X.Y.; Bone, A.J.; Brown, G.E.; Tanguay, R.L.; Di Giulio, R.T.; et al. Sulfidation of silver nanoparticles: Natural antidote to their toxicity. *Environ. Sci. Technol.* **2013**, *47*, 13440–13448. [CrossRef] [PubMed]

155. Petersen, E.J.; Diamond, S.A.; Kennedy, A.J.; Goss, G.G.; Ho, K.; Lead, J.; Hanna, S.K.; Hartmann, N.B.; Hund-Rinke, K.; Mader, B.; et al. Adapting OECD Aquatic Toxicity Tests for Use with Manufactured Nanomaterials: Key Issues and Consensus Recommendations. *Environ. Sci. Technol.* **2015**, *49*, 9532–9547. [CrossRef] [PubMed]
156. Corsi, I.; Fiorati, A.; Grassi, G.; Bartolozzi, I.; Daddi, T.; Melone, L.; Punta, C. Environmentally sustainable and ecosafe polysaccharide-based materials for water nano-treatment: An eco-design study. *Materials (Basel)* **2018**, *11*, 1228. [CrossRef]
157. Pedrazzo, A.R.; Smarra, A.; Caldera, F.; Musso, G.; Dhakar, N.K.; Cecone, C.; Hamedi, A.; Corsi, I.; Trotta, F. Eco-friendly ß-cyclodextrin and linecaps polymers for the removal of heavy metals. *Polymers (Basel)* **2019**, *11*, 1658. [CrossRef]
158. Fiorati, A.; Grassi, G.; Graziano, A.; Liberatori, G.; Pastori, N.; Melone, L.; Bonciani, L.; Pontorno, L.; Punta, C.; Corsi, I. Eco-design of nanostructured cellulose sponges for sea-water decontamination from heavy metal ions. *J. Clean. Prod.* **2020**, *246*. [CrossRef]
159. Holden, P.A.; Gardea-Torresdey, J.L.; Klaessig, F.; Turco, R.F.; Mortimer, M.; Hund-Rinke, K.; Cohen Hubal, E.A.; Avery, D.; Barceló, D.; Behra, R.; et al. Considerations of Environmentally Relevant Test Conditions for Improved Evaluation of Ecological Hazards of Engineered Nanomaterials. *Environ. Sci. Technol.* **2016**, *50*, 6124–6145. [CrossRef]
160. Liberatori, G.; Grassi, G.; Guidi, P.; Bernardeschi, M.; Fiorati, A.; Scarcelli, V.; Genovese, M.; Faleri, C.; Protano, G.; Frenzilli, G.; et al. Ecotoxicity assessment of environmental safety and efficacy of nanostructured cellulose sponges for zinc removal from seawater to prevent ecological risks. *Nanomaterials* **2020**, *10*, 1283. [CrossRef]
161. Alavi, M.; Rai, M. Recent progress in nanoformulations of silver nanoparticles with cellulose, chitosan, and alginic acid biopolymers for antibacterial applications. *Appl. Microbiol. Biotechnol.* **2019**, *103*, 8669–8676. [CrossRef]
162. Pivec, T.; Hribernik, S.; Kolar, M.; Kleinschek, K.S. Environmentally friendly procedure for in-situ coating of regenerated cellulose fibres with silver nanoparticles. *Carbohydr. Polym.* **2017**, *163*, 92–100. [CrossRef]
163. Jain, S.; Bhanjana, G.; Heydarifard, S.; Dilbaghi, N.; Nazhad, M.M.; Kumar, V.; Kim, K.H.; Kumar, S. Enhanced antibacterial profile of nanoparticle impregnated cellulose foam filter paper for drinking water filtration. *Carbohydr. Polym.* **2018**, *202*, 219–226. [CrossRef]
164. Morena, A.G.; Roncero, M.B.; Valenzuela, S.V.; Valls, C.; Vidal, T.; Pastor, F.I.J.; Diaz, P.; Martínez, J. Laccase/TEMPO-mediated bacterial cellulose functionalization: production of paper-silver nanoparticles composite with antimicrobial activity. *Cellulose* **2019**, *26*, 8655–8668. [CrossRef]
165. Kumar, V.; Pathak, P.; Bhardwaj, N.K. Waste paper: An underutilized but promising source for nanocellulose mining. *Waste Manag.* **2020**, *102*, 281–303. [CrossRef] [PubMed]
166. Kargarzadeh, H.; Ahmad, I.; Thomas, S. *Handbook of Cellulose Nanocomposites*; John Wiley & Sons, Incorporated, Wiley Online Library: Hoboken, NJ, USA, 2017.
167. Lee, M.; Oh, K.; Choi, H.K.; Lee, S.G.; Youn, H.J.; Lee, H.L.; Jeong, D.H. Subnanomolar Sensitivity of Filter Paper-Based SERS Sensor for Pesticide Detection by Hydrophobicity Change of Paper Surface. *ACS Sens.* **2018**, *3*, 151–159. [CrossRef] [PubMed]
168. Thomas, B.; Raj, M.C.; Athira, B.K.; Rubiyah, H.M.; Joy, J.; Moores, A.; Drisko, G.L.; Sanchez, C. Nanocellulose, a Versatile Green Platform: From Biosources to Materials and Their Applications. *Chem. Rev.* **2018**, *118*, 11575–11625. [CrossRef]
169. Isogai, A.; Saito, T.; Fukuzumi, H. TEMPO-oxidized cellulose nanofibers. *Nanoscale* **2011**, *3*, 71–85. [CrossRef]
170. Pierre, G.; Punta, C.; Delattre, C.; Melone, L.; Dubessay, P.; Fiorati, A.; Pastori, N.; Galante, Y.M.; Michaud, P. TEMPO-mediated oxidation of polysaccharides: An ongoing story. *Carbohydr. Polym.* **2017**, *165*, 71–85. [CrossRef]
171. Pääkko, M.; Ankerfors, M.; Kosonen, H.; Nykänen, A.; Ahola, S.; Österberg, M.; Ruokolainen, J.; Laine, J.; Larsson, P.T.; Ikkala, O.; et al. Enzymatic hydrolysis combined with mechanical shearing and high-pressure homogenization for nanoscale cellulose fibrils and strong gels. *Biomacromolecules* **2007**, *8*, 1934–1941. [CrossRef]
172. Tibolla, H.; Pelissari, F.M.; Menegalli, F.C. Cellulose nanofibers produced from banana peel by chemical and enzymatic treatment. *LWT Food Sci. Technol.* **2014**, *59*, 1311–1318. [CrossRef]

173. Bauli, C.R.; Rocha, D.B.; de Oliveira, S.A.; Rosa, D.S. Cellulose nanostructures from wood waste with low input consumption. *J. Clean. Prod.* **2019**, *211*, 408–416. [CrossRef]
174. Ngwabebhoh, F.A.; Yildiz, U. Nature-derived fibrous nanomaterial toward biomedicine and environmental remediation: Today's state and future prospects. *J. Appl. Polym. Sci.* **2019**, *136*. [CrossRef]
175. Wang, D. A critical review of cellulose-based nanomaterials for water purification in industrial processes. *Cellulose* **2019**, *26*, 687–701. [CrossRef]
176. Sehaqui, H.; Spera, P.; Huch, A.; Zimmermann, T. Nanoparticles capture on cellulose nanofiber depth filters. *Carbohydr. Polym.* **2018**, *201*, 482–489. [CrossRef] [PubMed]
177. Fiorati, A.; Turco, G.; Travan, A.; Caneva, E.; Pastori, N.; Cametti, M.; Punta, C.; Melone, L. Mechanical and drug release properties of sponges from cross-linked cellulose nanofibers. *Chempluschem* **2017**, *82*, 848–858. [CrossRef] [PubMed]
178. Melone, L.; Rossi, B.; Pastori, N.; Panzeri, W.; Mele, A.; Punta, C. TEMPO-Oxidized Cellulose Cross-Linked with Branched Polyethyleneimine: Nanostructured Adsorbent Sponges for Water Remediation. *Chempluschem* **2015**, *80*, 1408–1415. [CrossRef]
179. Ismail, M.; Khan, M.I.; Akhtar, K.; Seo, J.; Khan, M.A.; Asiri, A.M.; Khan, S.B. Phytosynthesis of silver nanoparticles; naked eye cellulose filter paper dual mechanism sensor for mercury ions and ammonia in aqueous solution. *J. Mater. Sci. Mater. Electron.* **2019**. [CrossRef]
180. Lebogang, L.; Bosigo, R.; Lefatshe, K.; Muiva, C. Ag_3PO_4/nanocellulose composite for effective sunlight driven photodegradation of organic dyes in wastewater. *Mater. Chem. Phys.* **2019**, *236*. [CrossRef]
181. Ali, A.; Mannan, A.; Hussain, I.; Hussain, I.; Zia, M. Effective removal of metal ions from aquous solution by silver and zinc nanoparticles functionalized cellulose: Isotherm, kinetics and statistical supposition of process. *Environ. Nanotechnol. Monit. Manag.* **2018**, *9*, 1–11. [CrossRef]
182. Ali, A.; Gul, A.; Mannan, A.; Zia, M. Efficient metal adsorption and microbial reduction from Rawal Lake wastewater using metal nanoparticle coated cotton. *Sci. Total Environ.* **2018**, *639*, 26–39. [CrossRef]
183. Goswami, M.; Baruah, D.; Das, A.M. Green synthesis of silver nanoparticles supported on cellulose and their catalytic application in the scavenging of organic dyes. *New J. Chem.* **2018**, *42*, 10868–10878. [CrossRef]
184. Zhang, Y.; Chen, L.; Hu, L.; Yan, Z. Characterization of Cellulose/Silver Nanocomposites Prepared by Vegetable Oil-Based Microemulsion Method and Their Catalytic Performance to 4-Nitrophenol Reduction. *J. Polym. Environ.* **2019**, *27*, 2943–2955. [CrossRef]
185. Dong, H.; Snyder, J.F.; Tran, D.T.; Leadore, J.L. Hydrogel, aerogel and film of cellulose nanofibrils functionalized with silver nanoparticles. *Carbohydr. Polym.* **2013**, *95*, 760–767. [CrossRef] [PubMed]
186. Pawcenis, D.; Chlebda, D.K.; Jędrzejczyk, R.J.; Leśniak, M.; Sitarz, M.; Łojewska, J. Preparation of silver nanoparticles using different fractions of TEMPO-oxidized nanocellulose. *Eur. Polym. J.* **2019**, *116*, 242–255. [CrossRef]
187. Chan, C.H.; Chia, C.H.; Zakaria, S.; Sajab, M.S.; Chin, S.X. Cellulose nanofibrils: A rapid adsorbent for the removal of methylene blue. *RSC Adv.* **2015**, *5*, 18204–18212. [CrossRef]
188. Han, Y.; Wu, X.; Zhang, X.; Zhou, Z.; Lu, C. Reductant-Free Synthesis of Silver Nanoparticles-Doped Cellulose Microgels for Catalyzing and Product Separation. *ACS Sustain. Chem. Eng.* **2016**, *4*, 6322–6331. [CrossRef]
189. An, X.; Long, Y.; Ni, Y. Cellulose nanocrystal/hexadecyltrimethylammonium bromide/silver nanoparticle composite as a catalyst for reduction of 4-nitrophenol. *Carbohydr. Polym.* **2017**, *156*, 253–258. [CrossRef]
190. Cruz-Tato, P.; Ortiz-Quiles, E.O.; Vega-Figueroa, K.; Santiago-Martoral, L.; Flynn, M.; Díaz-Vázquez, L.M.; Nicolau, E. Metalized Nanocellulose Composites as a Feasible Material for Membrane Supports: Design and Applications for Water Treatment. *Environ. Sci. Technol.* **2017**, *51*, 4585–4595. [CrossRef] [PubMed]
191. Tang, J.; Shi, Z.; Berry, R.M.; Tam, K.C. Mussel-inspired green metallization of silver nanoparticles on cellulose nanocrystals and their enhanced catalytic reduction of 4-nitrophenol in the presence of β-cyclodextrin. *Ind. Eng. Chem. Res.* **2015**, *54*, 3299–3308. [CrossRef]

© 2020 by the authors. Licensee MDPI, Basel, Switzerland. This article is an open access article distributed under the terms and conditions of the Creative Commons Attribution (CC BY) license (http://creativecommons.org/licenses/by/4.0/).

Article

Size-Controlled Transformation of Cu₂O into Zero Valent Copper within the Matrix of Anion Exchangers via Green Chemical Reduction

Irena Jacukowicz-Sobala [1,*], Ewa Stanisławska [1], Agnieszka Baszczuk [2], Marek Jasiorski [2] and Elżbieta Kociołek-Balawejder [1]

1. Department of Industrial Chemistry, Wrocław University of Economics and Business, ul. Komandorska 118/120, 53-345 Wrocław, Poland; ewa.stanislawska@ue.wroc.pl (E.S.); elzbieta.kociolek-balawejder@ue.wroc.pl (E.K.-B.)
2. Department of Mechanics, Materials Science and Engineering, Wrocław University of Science and Technology, ul. Smoluchowskiego 25, 50-370 Wrocław, Poland; agnieszka.baszczuk@pwr.edu.pl (A.B.); marek.jasiorski@pwr.edu.pl (M.J.)
* Correspondence: irena.jacukowicz@ue.wroc.pl

Received: 14 October 2020; Accepted: 5 November 2020; Published: 9 November 2020

Abstract: Composite materials containing zero valent copper (ZVC) dispersed in the matrix of two commercially available strongly basic anion exchangers with a macroreticular (Amberlite IRA 900Cl) and gel-like (Amberlite IRA 402OH) structure were obtained. Cu^0 particles appeared in the resin phase as the product of the reduction of the precursor, i.e., copper oxide(I) particles previously deposited in the two supporting materials. As a result of a one-step transformation of preformed Cu_2O particles as templates conducted using green reductant ascorbic acid and under mild conditions, macroporous and gel-type hybrid products containing ZVC were obtained with a total copper content of 7.7 and 5.3 wt%, respectively. X-ray diffraction and FTIR spectroscopy confirmed the successful transformation of the starting oxide particles into a metallic deposit. A scanning electron microscopy study showed that the morphology of the deposit is mainly influenced by the type of matrix exchanger. In turn, the drying steps were crucial to its porosity and mechanical resistance. Because both the shape and size of copper particles and the internal structure of the supporting solid materials can have a decisive impact on the potential applications of the obtained materials, the results presented here reveal a great possibility for the design and synthesis of functional nanocrystalline solids.

Keywords: zero valent copper; Cu^0-containing hybrid anion exchanger; Cu_2O reduction; ascorbic acid as reducer

1. Introduction

The development of nanoscience and nanotechnology in the past two decades has opened up new areas for the use of copper in the form of ultrafine particles with specific chemical, electrical, optical, and thermal properties dependent mainly on their size and shape. Because of their extremely small size, high surface energy, and large surface area/volume ratio, Cu nanoparticles (CuNPs) have different (e.g., physicochemical and biological) properties than those of their bulk metallic form [1–3]. Owing to their antimicrobial activity against a wide range of microorganisms, including multidrug-resistant organisms and fungi, CuNPs have been a strong focus of research on health-related processes [4,5]. They have also attracted considerable attention in the fields of heterogeneous catalytic reactions, including industrially important processes such as reduction of aromatic nitrocompounds, azide-alkyne cycloaddition, coupling reactions, advanced oxidation processes, and also photo- or electrocatalytic processes [2,3,6–10].

CuNPs have been synthesized using a variety of techniques. Among them, chemical reduction is the most frequently used due to its simplicity, limited equipment requirements, low cost, and possibility of synthesizing copper nano/microstructures with various morphologies. In these reactions, the source of copper most often is a cupric salt reduced by hydrazine or sodium borohydride. By selecting a reducing agent and varying the reaction conditions, one can control the size and shape of the obtained copper particles [11–15]. Unfortunately, the latter, due to van der Waals forces, tend to be rather unstable in solution, which leads to large agglomerates during the dispersion processing (preparation and storage). Moreover, copper nanoparticles are susceptible to oxidation when exposed to air, so the synthesis of stable products requires the presence of stabilizers. In order to protect CuNPs against their sticking and oxidation, surface-active agents, such as surfactants, dissolved polymers, and organic ligands (known as capping agents), can be used in aqueous media as dispersants and surface modifiers to yield highly stable, well-dispersed particles [13,16,17]. Another promising strategy concerning the stability of CuNPs is their immobilization on supporting solid materials, both of natural origin (cellulose, cotton, paper, chitosan, carbon active) and synthetic polymers (zeolite, graphene oxide, cation exchangers) [18–23].

The group of polymeric/inorganic composites based on ion exchangeable reactive polymers, also called hybrid ion exchangers (HIXs), is significantly prospective due to the presence of different active components such as a metal deposit and positively or negatively charged functional groups of supporting polymers [24–28]. The dispersion of NPs within the matrix of such a host material not only prevents their agglomeration and ensures contact between the reagents, but it also, owing to many possible combinations of hybrid polymers' components, enables one to develop multifunctional materials, which frequently exhibit synergetic effects. The physical form of spherical beads of HIXs makes them suitable (as opposed to parent NPs) for use in dynamic conditions in fixed bed column systems. In turn, the functional groups of the polymer modify their activity by the preconcentration of the reagents in the solid phase of the HIXs. Furthermore, functional groups may also have specific catalytic, stabilizing, or biocidal properties (in addition to CuNP activity) [1,29–31].

Copper-containing hybrid ion exchangers are synthesized based on sulfonic and carboxylic cation exchangers useful as functional materials for water deoxygenation. The introduction of metallic copper into cation exchangers containing negatively charged functional groups (sulfonic, carboxylic) is not difficult because these groups in the ion exchange reaction quantitatively bind Cu^{2+} ions from the solution. When a cation exchanger in the Cu^{2+} form was treated with a reducing agent solution (hydrazine, sodium dithionite), Cu^0 particles precipitated in its matrix [32–34]. Considering the multifunctionality of the hybrid polymers containing copper deposit, the choice of an anion exchanger as a supporting material is a very promising challenge. This choice is particularly interesting due to the presence of quaternary ammonium groups, which not only may be involved in preconcentration of negatively charged reagents in the solid phase of the ion exchanger but also show antimicrobial activity (in contrast to the functional groups of the cation exchangers) and can enhance the biocidal efficiency of the obtained CuNP-containing composites [35]. However, it is impossible to introduce Cu^0 particles into the anion exchange matrix in an analogous way to the synthesis of Cu^0-containing cation exchangers. The positively charged quaternary ammonium functional groups repel Cu^{2+} ions. Therefore, using the anion exchanger as a supporting material is challenging and requires an unconventional strategy.

In considering ways of obtaining anion exchangers with metallic copper, we came to the conclusion that this aim could be achieved most easily using anion exchangers containing Cu_2O, as described in our previous studies [36]. This material was selected as a candidate for the production of HIXs containing metallic copper because it contains low oxidation copper atoms, which can be relatively easily reduced at room temperature with ascorbic acid [15,37–40]. Strong reducers, such as hydrazine and $NaBH_4$, are not only toxic and flammable chemicals, but they also release a gaseous product in the reaction, which could be unbeneficial when the substrate subjected to the reaction is an anion exchanger in the form of a porous solid. Low temperature during reduction is necessary due to low thermal resistance of anion exchangers (up to 60–70 °C).

Therefore, the aim of this study is to transform the Cu_2O deposited in the anion exchanger's matrix to Cu^0 through a simple and environmentally benign reaction with ascorbic acid as the reducer. More precisely, our objectives are (1) to determine whether the transformation of Cu_2O to Cu^0, which easily proceeds in (bulk) aqueous media, would also occur in mild conditions in the solid phase of anion exchangers; (2) to examine the influence of reaction conditions such as pH < 7, time, temperature, and intermediate drying on the efficiency of Cu_2O conversion into Cu^0 within the polymeric matrix; and (3) to determine whether the type of supporting polymer structure (macroporous or gel-type) influences the distribution, size, and shape of deposited CuNPs. The present study is unique since its aim is not simply to introduce a deposit into the anion exchanger phase but to determine how as a result of a chemical reaction, the introduced deposit changes into a deposit characterized by a different chemical structure (and specific physicochemical properties). Such composite materials—zero valent copper (ZVC) immobilized and stabilized in the matrix of anion exchangers (An/Cu)—deserve special attention due to their unique method of synthesis, unexpected physicochemical properties, and potential applicability in multidisciplinary fields connected with environmental protection.

2. Materials and Methods

2.1. Materials

The polymer supports for Cu_2O and Cu^0 were Amberlite IRA 900Cl (M/An) and Amberlite IRA 402OH (G/An)—two commercially available anion exchange resins produced by The Dow Chemical Co. (Midland, MA, USA). All the chemicals used in this study, including ascorbic acid(+) (POCh Gliwice, Poland), $CuCl_2·2H_2O$ (Chempur, Piekary Śląskie, Poland), HNO_3 65% (PPH Stanlab, Lublin, Poland), NaOH, HCl 35%, and cuprizone 98% (Alfa Aesar, (Ward Hill, MA, USA)), were of analytical grade. All the solutions were prepared using deionized water.

2.2. Synthesis of An/Cu_2O

The starting anion exchangers were dried at 40 °C for 24 h in a dryer chamber. To transform the functional groups in the $CuCl_4^{2-}$ form, a sample of the anion exchanger weighing about 2.0 g was placed in a conical flask and treated with 20 cm^3 of 0.5 mol dm^{-3} $CuCl_2$ in 5 mol dm^{-3} HCl. The reagents were shaken at 20 °C for 1 h (M/An) or for 4 h (G/An). To carry out the reduction reaction, the filtered off (under vacuum) intermediate product was placed in a conical flask, was treated with 40 cm^3 of the 0.5 mol dm^{-3} ascorbic acid (AA) solution in 2 mol dm^{-3} NaOH, and was shaken at 20 °C for 24 h. After this time, the sediment was removed from the reaction medium through decantation, and the filtered off and washed product (An/Cu_2O) was dried at 40 °C for 24 h or kept in the wet state [36].

2.3. Synthesis of An/Cu^0

The synthesis of An/Cu^0 was conducted using two methods depending on the form of starting material An/Cu_2O—i.e., dried at 40 °C (Method 1) or wet state filtered off under vacuum to moisture content 50 % (Method 2). A sample of An/Cu_2O containing about 0.5 g dry matter was placed in a conical flash and treated with 20 cm^3 of the 1 mol dm^{-3} AA solution in water alone or in 1 mol dm^{-3} NaOH (Table 1). The reagents were shaken at 20 °C for 24 h or at 50 °C for 3 h (or for 30 min). After this time, any sediment was removed from the reaction medium through decantation, and the filtered off and washed product (An/Cu) was dried at 40 °C for 24 h.

Table 1. Reducer solutions.

Symbol of Solution	Reducing Agent Solution	Potential (mV)	pH
AA/2	1 M ascorbic acid	+113	2.12
AA/6	1 M ascorbic acid in 1 M NaOH	−195	6.42

2.4. Characterization

The total content of Cu in An/Cu was determined using inductively coupled plasma atomic emission spectroscopy (Thermo Scientific iCAP 7400, Waltham, MA, USA) after a sample mineralization using the digestion microwave system MARS 5 (CEM Corporation, Matthews, NC, USA). The Cu content in the An/Cu was determined also after dissolving Cu^0 in nitric acid solution. A sample of An/Cu weighing about 0.2 g was treated with 10 cm^3 of HNO_3 (1:3) in a conical flask and shaken for 24 h. After this time, the polymer sample was filtered off and washed with deionized water. The solution was analyzed spectrophotometrically using the cuprizone (bis(cyclohexanone)oxaldihydrazone) method. Analyses were repeated twice, and the standard deviation did not exceed 1.0%. The redox potential values of the reducer solutions were measured with the electrode ERPt-11X1 (Hydromet, Gliwice, Poland).

The synthesized materials were investigated by scanning electron microscopy (SEM), X-ray diffraction, and FTIR spectroscopy. The morphology and elemental composition of the different An/Cu samples were examined with a Hitachi S-3400 N scanning electron microscope (Tokyo, Japan) equipped with an energy-dispersive spectrometry (EDS) microanalyzer. The FTIR spectra of prepared An/Cu samples were measured using a Fourier transform Bruker VERTEX 70 V vacuum spectrometer equipped with an air-cooled DTGS detector (Billerica, MA, USA). The XRD patterns were measured on an X-ray Rigaku Ultima IV diffractometer using a Cu Kα radiation source (λ = 1.5406 Å) operating at 40 kW/40 mA with the range from 10° to 100° in 0.05° steps with an exposure time of 4 s per point.

To analyze the surface and pore parameters of the M/An/Cu and M/An/Cu-2, the vacuum filtered off product, after washing with deionized water, was dried by applying thermally drying or freeze-drying techniques. The freeze-drying process was conducted by treating the materials at −80°C for 24 h, followed by drying at a pressure of 0.02 mbar for 24 h using a Labconco FreeZone Laboratory Freeze Dryer 4.5 L. The porous characteristics of the samples were determined by the low temperature N_2 adsorption–desorption at 77 K using an ASAP 2020 Micromeritics porosimetry analyzer; mercury intrusion porosimetry was conducted on a Micrometrics AutoPore IV 9510 analyzer.

3. Results and Discussion

3.1. Optimization of ZVC Content within the Matrix of Anion Exchangers

The chemical deposition of the Cu^0 particles inside a strongly basic anion exchanger represents a multistage process that has three steps (Scheme 1).

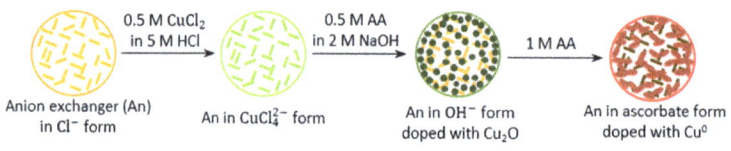

Scheme 1. Rout of synthesis of hybrid ion exchangers (HIXs) containing metallic copper deposit.

In order to accomplish the last transformation and obtain an anion exchanger with metallic copper particles in the resin phase, the following reaction (1) was carried out:

$$\text{[ascorbic acid]} + Cu_2O \longrightarrow \text{[dehydroascorbic acid]} + 2Cu + H_2O \quad (1)$$

Two commercially available anion exchangers containing the same type and comparable number of strongly basic quaternary ammonium functional groups (3.0 meq g^{-1}), but varying in the structure of the poly(styrene/divinylbenzene) skeleton (macroporous Amberlite IRA 900Cl or gel-like Amberlite IRA402OH), were used as the polymeric support for the ZVC. The Cu$_2$O deposit was introduced into the anion exchanger matrix in two steps: First, the anion exchanger functional groups were transformed into the CuCl$_4^{2-}$ form, and then the intermediate polymeric product was contacted with an alkaline ascorbic acid solution [36]. It was this reducer that in a highly alkaline environment made it possible to practically quantitatively precipitate Cu$_2$O into the resin phase (not into the outside of the latter as in the case of glucose). In our previous study, it was shown that a large amount of this deposit could be successfully introduced into both macroreticular and gel-like anion exchangers. The very high Cu$_2$O content and the advantageous distribution of the deposit throughout the volume of ion exchanger beads (Table 2) make them ideal substrates for synthesizing even more unique materials, such as metallic copper-doped anion exchangers. Detailed morphological characteristics were described in our previous study [36].

Table 2. Hybrid ion exchangers with Cu$_2$O deposit.

Properties	Symbol of HIX	
	M/An/Cu$_2$O	G/An/Cu$_2$O
Polymeric carrier	Amberlite IRA 900Cl	Amberlite IRA 402OH
Physical form	Spherical beads	Spherical beads
Polymer matrix	Styrene/divinylbenzene copolymer	Styrene/divinylbenzene copolymer
Functional groups	Trimethylammonium	Trimethylammonium
Matrix structure	Macroreticular	Gel-like
Inorganic deposit	Cu$_2$O	Cu$_2$O
Cu content in HIX, wt%	8.6	6.6
Appearance of dispersed Cu$_2$O deposit	Unsymmetrical bracelet-like clusters of a few to 10–20 adjoining spherical objects approximately 200 nm in diameter	Highly symmetrical, single, separate, ideally shaped spheres approximately 1.0 µm in diameter

It was reported in the literature that by reducing cupric compounds with ascorbic acid, one can obtain different products, depending on the reaction medium pH. If ZVC is to be obtained, pH < 7.0 should be ensured. Moreover, it has been shown that the lower the pH, the faster Cu^0 particles form, and the higher the pH, the smaller are the particles [37–40]. Considering that the reactions described in the literature were conducted in bulk solutions in different conditions, in order to transform Cu$_2$O to Cu^0 in the anion exchanger matrix (i.e., in atypical conditions this is more difficult because of the necessity of diffusing the reagents within the polymeric skeleton of the supporting material), first, the experiments whose results are presented as Method 1 in Table 3 were carried out. Two reaction solutions with pH < 7, but differing in their acidity—namely AA/2, a solution of ascorbic acid alone (with pH ~2), and AA/6, a solution neutralized with an equimolar amount of NaOH (sodium ascorbate, with pH ~6)—were used, and an appropriately long time of contact between the reagents at different temperatures was ensured.

Table 3. Conditions and results of Cu_2O reduction.

Symbol of HIX	Solution	Temp. (°C)	Reaction Time (h)	Cu Content in An/Cu (wt%)
Method 1, dry Cu_2O as a precursor of Cu				
M/An/Cu_2O	AA/2	20	24	6.76
		50	3	7.68
	AA/6	20	24	6.65
		50	3	6.83
G/An/Cu_2O	AA/2	20	24	4.53
		50	3	4.50
	AA/6	20	24	5.35
		50	3	5.32
Method 2, wet Cu_2O as a precursor of Cu				
M/An/Cu_2O	AA/2	50	0.5	6.90
G/An/Cu_2O				5.10

The data in Table 3 indicate that regardless of the reduction conditions, after the reaction all the products contained much copper. This means that the skeletons of both the macroreticular and gel-like anion exchangers retained the deposit undergoing transformation from Cu_2O to Cu^0. Even though the Cu content in the An/Cu matrix, amounting to 4.50–7.68%, expressed the actual copper content in the sample, this result did not fully reflect the degree of transformation of Cu_2O into Cu^0. The mass of the An/Cu samples was found to be up to 25% greater than that of the initial An/Cu_2O samples, which was owing to the transformation of the anion exchanger's functional groups from the hydroxyl into the ascorbate form. It is worth emphasizing here that this form of exchanger functional groups is particularly advantageous due to its antioxidant properties protecting the newly formed copper from undesirable oxidation. Considering the gain in sample mass, one can estimate that in the case of the samples with the highest copper content, the degree of conversion amounted to over 95%, and in none of the cases was it lower than 85%. Thus, it turned out that neither the long shaking of the sample nor the "loosening" of the polymer structure at the elevated temperature nor the character of the reaction medium (ascorbic acid with a low pH or an almost neutral sodium ascorbate solution) resulted in any loss of a significant part of the inorganic deposit, which is difficult to achieve when attempting to obtain such materials as HIXs. However, it should be noted that when M/An/Cu_2O was the reagent, the reaction medium became slightly cloudy, which was ascribed to the crumbling of the ion exchanger rather than to the passing of the inorganic deposit to the aqueous phase.

3.2. Characterization of the Form of Inorganic Deposit

3.2.1. XRD Analysis

The content of Cu deposited in the structure of anion exchangers (Table 3) may not totally reflect the content of zero valent copper nanoparticles present in the samples determined after their mineralization. Since both the precursor (Cu_2O) and product (metallic Cu particles) form crystalline structures, eight samples of hybrid polymers obtained in different conditions were examined by X-ray diffraction analysis. XRD patterns obtained for hybrid anion exchanger samples are shown in Figure 1. The peaks visible on diffractograms of both types of ion exchangers, located at 43.3°, 50.4°, 74.2°, and 90.0° 2Θ, undoubtedly correspond to metallic copper (ICSD collection code: 64699) in the form of a single face-centered cubic cell. A closer analysis of Figure 1 reveals additional low-intensity reflections on the diffraction pattern of the gel HIX sample obtained after the reduction of G/An/Cu_2O at an elevated pH and temperature (pH ~6, 50 °C). The most intense of the additional peaks is located at 36.4° 2Θ, which indicates that it is associated with the presence of Cu_2O in the G/An/Cu sample. The presence of residual, unreduced Cu_2O is probably a consequence of too short reaction time (3 h) in spite of elevated temperature. In contrast, a 3 h duration of the process was sufficient for the reduction of Cu_2O in the structure of the macroporous anion exchanger due to a significantly better developed porous structure.

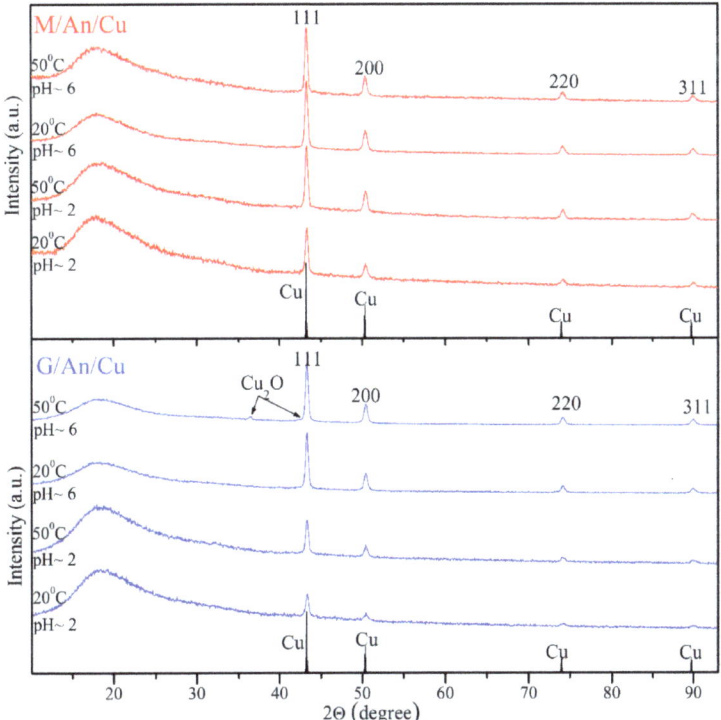

Figure 1. XRD patterns of reduction reaction products M/An/Cu and G/An/Cu.

As can be seen in Figure 1, in addition to distinct peaks from metallic copper, all diffractograms show an amorphous halo with a maximum at about 18°. The presence of this characteristic very wide peak can be associated with the amorphous polymer anion exchanger [41]. A comparison of diffractograms of macroreticular M/An/Cu samples obtained in different reduction conditions reveals that the ratio of the intensity of the broad polymer peak to the (111) peak of metallic copper is the highest ($\frac{I_{amorf}}{I_{(111)Cu}} > 1$) for the sample obtained at low pH and room temperature. In the remaining M/An/Cu samples, the (111) copper peak has a much higher intensity than that observed for the polymer ($\frac{I_{amorf}}{I_{(111)Cu}} < 1$), which indicates a significant content of metallic copper in these samples. An analysis of the XRD results of the gel G/An/Cu samples shows that the ratio of the intensity of the broad halo-called amorphous peak to the most intense (111) peak of metallic copper is always higher ($\frac{I_{amorf}}{I_{(111)Cu}} > 1$) in samples obtained at lower pH than for samples reduced at pH ~6 ($\frac{I_{amorf}}{I_{(111)Cu}} < 1$), regardless of the temperature during the reduction. This indicates a higher content of metallic deposit, regardless of the type of exchanger, in samples reduced at pH ~6. These results are compliant with the total Cu content in the studied hybrid polymers. It may be due to a higher dissociation degree of ascorbic acid at neutral pH (pK_a = 4.05) and its electrostatic attraction by positively charged functional groups of the polymer that the diffusion of the reducer into the matrix of anion exchanger was facilitated. In addition, the increase of pH decreases the redox potential of ascorbic acid, which makes it a more powerful reducer at higher pH [36].

3.2.2. FTIR Analysis

In order to further investigate the potential presence of other noncrystalline by-products, hybrid polymers with the highest copper content were analyzed by FTIR spectroscopy. Since Cu is inactive to IR analysis, we can only observe the peaks that can be assigned to the copper compounds, polymeric matrix, and counterions present in the functional groups of the polymer (Figure 2). In the FTIR spectrum, there appear numerous peaks which are indicative for the supporting anion exchanger, whose assignments are presented in Table 4. In accordance with our previous study [36], this analytical technique is not sufficient to strictly identify the copper oxide deposited in the matrix of the hygroscopic polymer, since their characteristic peaks indicative for Cu–O vibrations occur in the range of 500–630 cm^{-1}, which is also the frequency of a broad set of the absorption bands of hygroscopic water. However, as a consequence of the reduction of Cu_2O to Cu, in the spectrum of ZVC-containing hybrid polymer, a characteristic peak of Cu(I)–O vibrations at 630 cm^{-1} disappeared. Although FTIR spectra of the obtained hybrid materials represent the superposition of all components, it is possible to observe bands associated with ascorbic acid—the strong absorption in the range 1710–1775 cm^{-1} attributed to C=O stretching of the lactone ring (Figure 2a,b) [42]. Since the reaction occurs under the excess of ascorbic acid, its dissociated forms are bound as counter ions with the functional groups of the supporting anion exchanger. The presence of ascorbate groups manifested in IR spectra is very beneficial due to their strong antioxidant properties ensuring stability to copper metallic particles during storage [17,43,44]. It can also be observed that after reduction of Cu_2O into Cu^0, the area of a broad peak indicative of adsorbed water at 3350 cm^{-1} decreases and almost vanishes in the case of gel-type Cu^0-containing hybrid polymer. This relation suggests that after the conversion of copper oxide into zero valent copper, the deposit makes the surface of the product less hydrophilic, which is probably due to low chemical activity for water dissociation of the Cu surface, resulting in weak water adsorption properties [45].

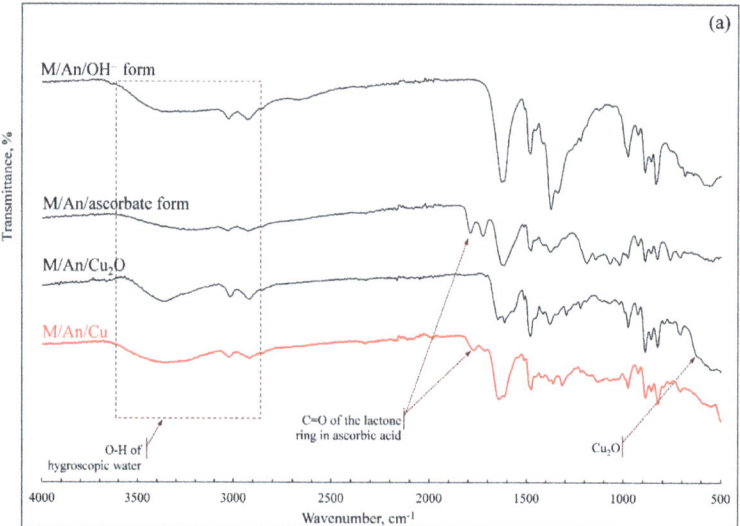

Figure 2. *Cont.*

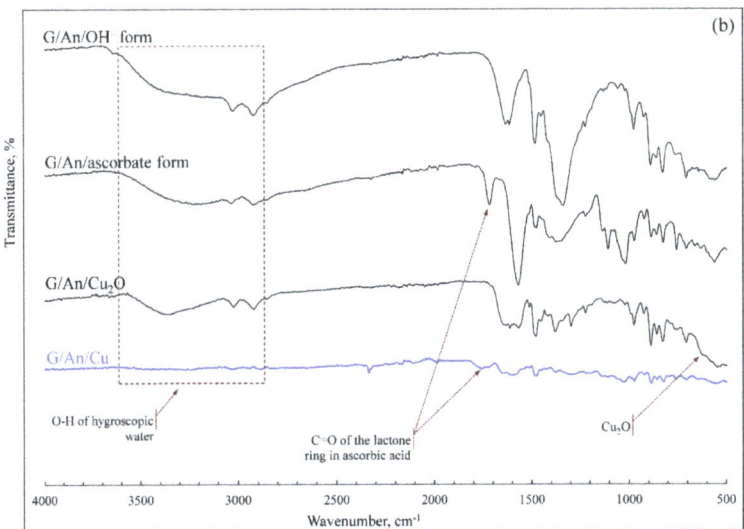

Figure 2. FTIR spectra of virgin anion exchangers in the OH⁻ and ascorbate form: M/An (**a**) and G/An (**b**); starting hybrid polymers containing Cu_2O: M/An/Cu_2O (**a**) and G/An/Cu_2O (**b**); and final products containing Cu^0 deposit: M/An/Cu (**a**) and G/An/Cu (**b**).

Table 4. Assignments of infrared bands of the host anion exchangers [46].

Assignment	Wavenumber cm^{-1}
O–H vibrations of hygroscopic water	1650, 3100–3600
C=C and C–H vibrations of benzene rings and aliphatic groups	1020, 1111, 1418, 1478, 1512, 1613, 3020
C–H, C–N, CCN vibrations in CH_3–N	707, 826, 859, 889, 926, 975, 1221, 1341, 1383, 2855, 2922

3.2.3. Scanning Electron Microscopy Studies

The morphology of the obtained hybrid polymers containing ZVC was studied using scanning electron microscopy. One sample for each type of supporting material, i.e., macroporous and gel-type anion exchanger showing the highest total copper content (in accordance with Table 3), was selected. Thus, among the samples of macroreticular exchangers, the M/An/Cu one obtained at higher temperature and using ascorbic acid with a low pH was tested and for gel exchangers those that were produced at room temperature with almost neutral reducer solution. Figure 3a shows that macroreticular grains are mostly broken. Their external surface is characterized by extensive cracks. Occasionally light spots appear on the surface (Figure 3b–d). According to the EDX (Energy Dispersive X-Ray Analysis), these spot areas contain only copper (Figure 3e). The morphology of these spots differs significantly from the morphology observed for other copper-containing objects. There are no visible particles on the spots but an almost solid layer, which suggests that these areas may be associated with amorphous copper. An EDX analysis, performed on the macroreticular bead cross section, revealed that the copper deposit is distributed throughout the volume of the M/An/Cu grain (Figure 3f). A more careful observation of the interior of the grain reveals a few clusters of closely lying individual particles with a size up to maximum 200 nm (Figure 3g,h).

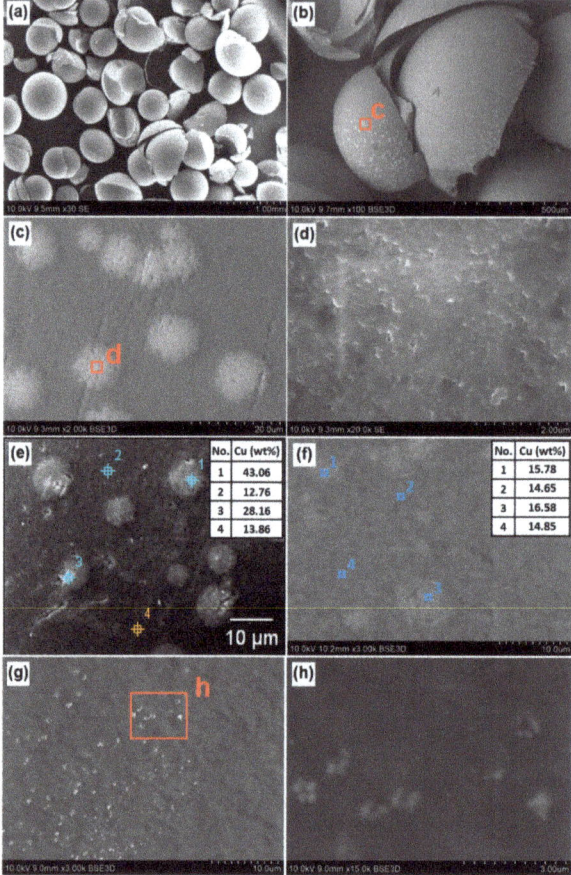

Figure 3. SEM images and EDS analysis of the M/An/Cu (**a–e**) surface, (**f–h**) inside.

Figure 4 shows the morphology of both the surface (Figure 4a–e) and the interior of the G/An/Cu beads (Figure 4f–h). Gel beads are evidently not broken, but their outer surface is clearly cracked. Extensive cracks are visible on virtually every grain (Figure 4b). In addition, a mesh of minor cracks is observed on some of them (Figure 4c). Higher fracture magnifications reveal numerous copper particles with characteristic hemispherical particle shape and sizes from objects smaller than 100 nm to a maximum of ~500 nm (Figure 4d,e). The interior of the grains is smooth and nonporous with numerous evenly distributed copper aggregates, with a total size of approximately 1 µm, formed from tightly packed particles about 200 nm in size (Figure 4f–h).

Comparing the morphology of the copper deposits obtained in the M/An/Cu and G/An/Cu anion exchangers, it can be seen that the internal structure of the anion exchanger has a decisive impact on the shape and size of the deposit. Our previous research on HIX containing Cu_2O, used here as a source for the growth of the Cu crystals, also led to similar conclusions [36]. Macroporous beads have permanent pores, in contrast to gel-type grains. Contrary to the pores in G/An, which appear only in the swollen state, the permanent internal macropores in M/An do not collapse when the material loses water, and for this reason they do not lead to dense packing of deposited particles. For this reason, in the M/An matrix no such dense packing of deposited particles was observed as in gel exchangers. In addition to the type of exchanger (its internal structure), the morphology of the Cu_2O precursor particles undoubtedly had a significant impact on the size and shape of the produced copper metallic particles. In the case of substrate macroreticular M/An/Cu$_2$O beads, the Cu$_2$O deposit was in the form of evenly dispersed, irregular clusters of a few to 10–20 adjoining spherical objects approximately 200 nm in diameter. After reductive conversion, the resulting copper particles mostly are not visible on SEM images due to their much smaller (compared to the original Cu$_2$O) size. Hardly visible copper-containing objects are in the form of clusters of a few particles up to a maximum of 200 nm. In the case of the substrate gel HIX (i.e., G/An/Cu$_2$O), this type of matrix induced the formation of compact Cu$_2$O aggregates of approximately 1 µm. After reductive conversion, the produced copper particles (about 200 nm in size) no longer form a compact aggregate but rather are loosely connected into a cluster with a diameter of about 1 µm. Apparently, visible changes in the size of Cu^0 aggregates that were formed after reduction result from a smaller molar volume of copper (7.61 cm^3/mol) compared to Cu$_2$O (23.85 cm^3/mol) [47]. However, the basic shape of the aggregates formed by nanoparticles is maintained after the reduction process because the initial Cu$_2$O nanoparticles are used as a template of transformation for the crystal-to-crystal conversion.

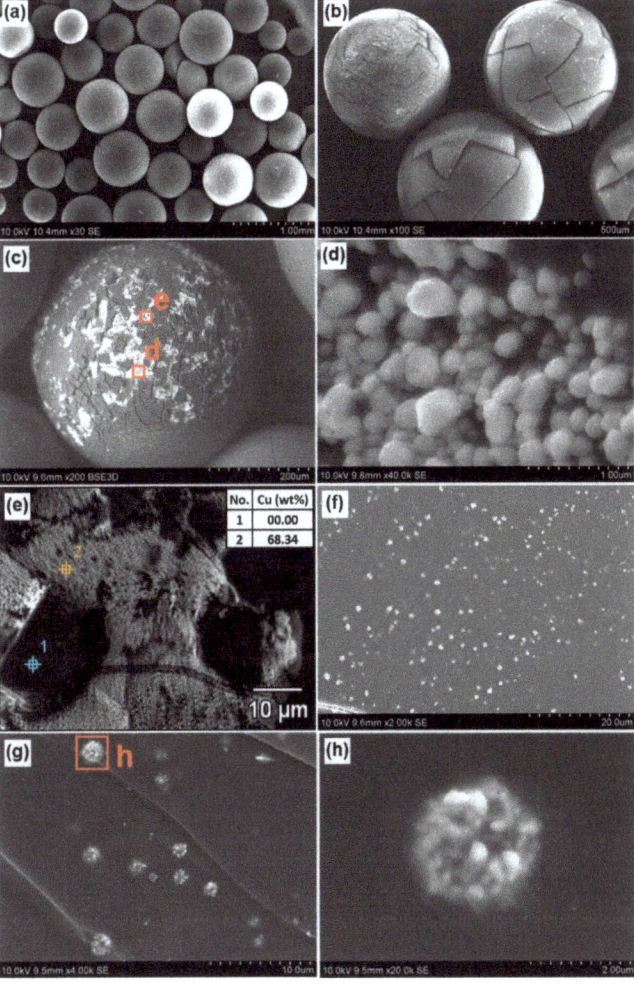

Figure 4. SEM images and EDS analysis of the G/An/Cu (**a**–**e**) surface, (**f**–**h**) inside.

3.2.4. Studies of Porosity

The deposition of ZVC load in the multistep process including drying and wetting in aqueous solutions by the shrink-swell mechanism may affect not only the physical integrity of the beads but also their porous structure. As can be seen in Figure 3a, showing the products obtained by pouring a dry An/Cu$_2$O sample into the solution, due to the rapid swelling of the ion exchanger, the beads, particularly the M/An/Cu ones, are cracked. This observation is also reflected in changes of macroporous structure of the thermally dried polymeric beads after Cu^0 deposition. A comparison of pore size distribution determined using the mercury intrusion method for the thermally dried product (M/An/Cu) with those obtained for the thermally dried initial anion exchanger (M/An) [48] shows that after ZVC incorporation, macropores with a diameter exceeding 1 µm appeared, and the pore size distribution changed from uniform to wide-ranging dimensions. This change is probably a result of resin structure disintegration, since aggregates of Cu^0 deposited within pores of the supporting material may influence its resistance to osmotic pressure [49]. A further analysis of porous parameters (Table 5)—BET (Brunauer, Emmett and Teller) surface area (micro and mesoporous structure) and porosity determined by mercury intrusion (macro- and mesoporous structure)—which decreased significantly after Cu^0 deposition from 21.7 m^2/g to almost 0.0 m^2/g and from 29.8 to 4.7%, respectively, confirm the blockage and closure of porous structure of the host polymer with aggregates of the inorganic constituent. Additionally, the effect of pore blockage may be amplified by copper aggregation and its more hydrophobic nature, resulting in less hygroscopic water content and a less swollen structure of the polymer, which is in agreement with the results of the FTIR analysis. In consequence, lower hydration, screening of pore walls, squeezing, and structure defects due to the presence of the inorganic deposit may lead to the shrinkage and collapse of the structure of the transport macropores, which is reflected in the more significant loss in the porosity and increase in apparent density. Simultaneously, the shape of the N$_2$ adsorption–desorption isotherm—type IV—and its hysteresis loop—type H1—of all studied materials is indicative for the presence of cylindrical mesopores, even in the case of thermally dried M/An/Cu, which showed a very small BET surface area (Figure 5a,b). This supports our conclusion that Cu^0 aggregates completely close the greater part of mesopores in the matrix of the dried anion exchanger. Since the gel-type ion exchangers in a dry state do not possess any internal surface area, the porous structure of gel-type products was not studied. During drying, their structure shrinks and the plot of the isotherm is different—it is a straight line with no hysteresis loop—indicating that the N$_2$ adsorption occurs only on the outer surface of the beads.

Table 5. Porous characteristics of supporting anion exchanger and Cu^0-doped anion exchanger (additional sample specification FD means freeze-dried samples).

Sample Code	S_{BET} [a] (m^2/g)	V_T (Pore Volume) [a] (cm^3/g)	L_m (Pore Diameter) [a] (nm)	Total Surface Area [b] (m^2/g)	Total Intrusion Volume [b] (cm^3/g)	Porosity [b] (%)	Bulk Density (g/cm^3)
			Amberlite IRA 900Cl *				
M/An	21.7	0.0544	5.0	60.95	0.3945	29.84	0.877
M/An/FD	27.8	0.0607	4.4	56.16	0.5195	37.17	0.716
			Product of Method 1				
M/An/Cu	0.13	0.001	Indeterminable	16.3	0.0364	4.68	1.288
M/An/Cu/FD	20.8	0.040	7.7	46.2	0.443	37.3	0.842
			Product of Method 2				
M/An/Cu-2/FD	18.4	0.037	7.9	43.5	0.390	34.6	0.886

[a] N$_2$ adsorption/desorption measurement, [b] mercury intrusion porosimetry, * results from previous studies [48].

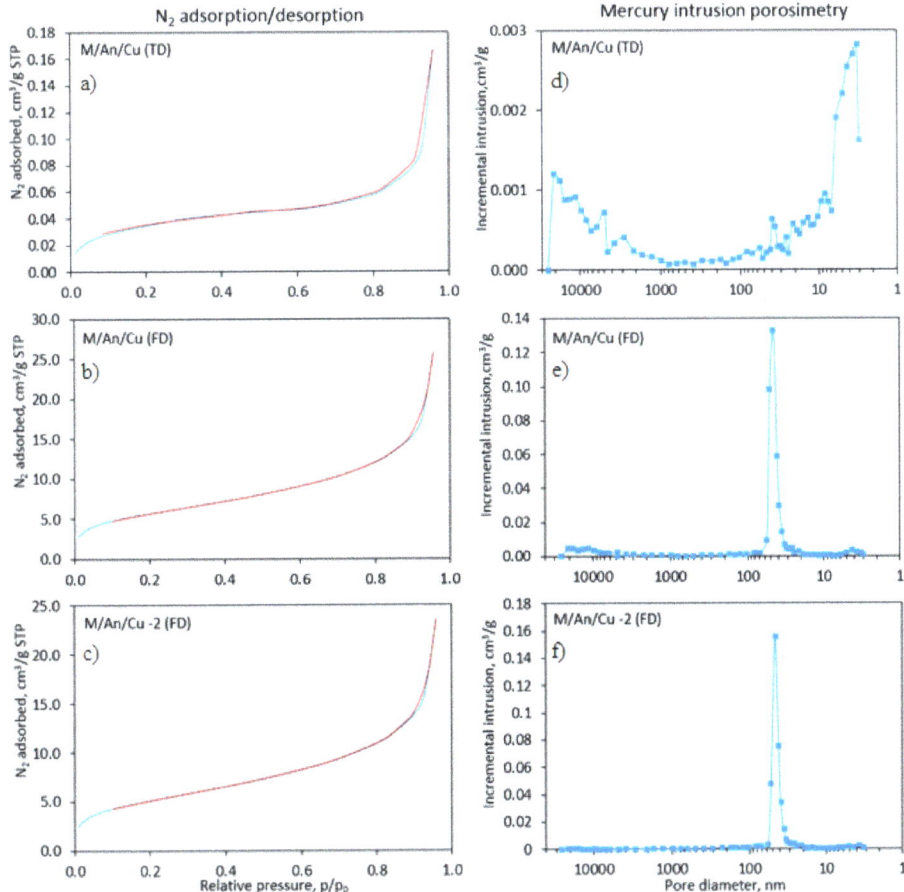

Figure 5. Isotherms of N_2 adsorption/desorption (**a–c**) and pore size distribution determined by mercury intrusion porosimetry (**d–f**) for thermally dried M/An/Cu (**a,d**) freeze-dried M/An/Cu (**b,e**) and freeze-dried M/An/Cu-2 obtained by method-2.

Unlike thermal drying, freeze drying of the hybrid anion exchangers allowed their porous structure to be preserved even in a dried state. The macroporosity did not change after Cu^0 deposition (Table 5), and the pore size distribution that was determined by mercury intrusion was very narrow, indicating the presence of mesopores with the maximum at 40 nm in the case of both the virgin supporting polymer [48,50] and the ZVC-containing hybrid product (Figure 5e,f). After Cu deposition, we observed a small volume increase in the range of macropores with dimensions exceeding 5 μm, which may be an effect of insignificant disintegration of the polymeric beads. The mesoporous structure of the beads was more significantly affected. The mean pore diameter determined from the N_2 adsorption isotherm increased from 4.4 to 7.7 nm while the BET surface area decreased from 27.8 m^2/g to 20.8 m^2/g, which probably resulted from the blockage of the smallest pores of the anion exchanger. Summing up, the method of drying is crucial to the mechanical resistance of the obtained materials while the changes in porous characteristics may be reversible due to the elastic structure of the polymeric matrix in aqueous solutions (by swelling in electrolyte solution and the subsequent reopening of the porous polymeric structure). Nevertheless, freeze drying seems to be a significantly better method when the application of the obtained material involves gaseous reagents. In turn, thermal drying resulting in the collapse of porous structure may restrict gas permeation and consequently the oxidation of ZVC load during storage in atmospheric air.

3.3. Operational Control of Physical Form and Copper Deposit Crystallinity of Hybrid Anion Exchangers

In order to avoid the shrinkage and swelling of the polymeric carrier, the synthesis was modified by introducing a vacuum-filtered (wet) An/Cu$_2$O sample into the reaction medium (Table 3, Method 2). Taking into consideration the previously established fact that the reaction rate is higher in the AA solution alone, only this solution and a shorter reaction time were used in the next stage of the research. Almost immediately after they had been introduced into the reaction medium, both the samples changed their color to claret, no sediment was observed in the aqueous phase, and the reaction products had uniform, undamaged grains (see the photos in Figure 6).

Figure 6. Photographs of examined materials: based on macroporous anion exchanger (left column) and gel-type anion exchanger (right column).

An XRD analysis (Figure 7) confirmed the assumption that the short reaction time (30 min) was sufficient in order to convert Cu$_2$O into Cu^0 in both the macroreticular and gel-like anion exchanger. In the case of the latter product, this result evidently showed that the swelling of the polymeric matrix significantly influences the kinetics of the reduction process (Table 3). A comparison of XRD diagrams of samples obtained by Method 1 (Figure 1, second line from the top in M/An/Cu and G/An/Cu part), with diffractograms of samples from Method 2 (Figure 7, M/An/Cu-2 and G/An/Cu-2) shows that the $\frac{I_{amorf}}{I_{(111)Cu}}$ ratios in the diagrams from Method 2 are clearly greater than in the case of the corresponding samples obtained by Method 1, which indicates a reduced fraction of crystalline Cu^0 in these samples.

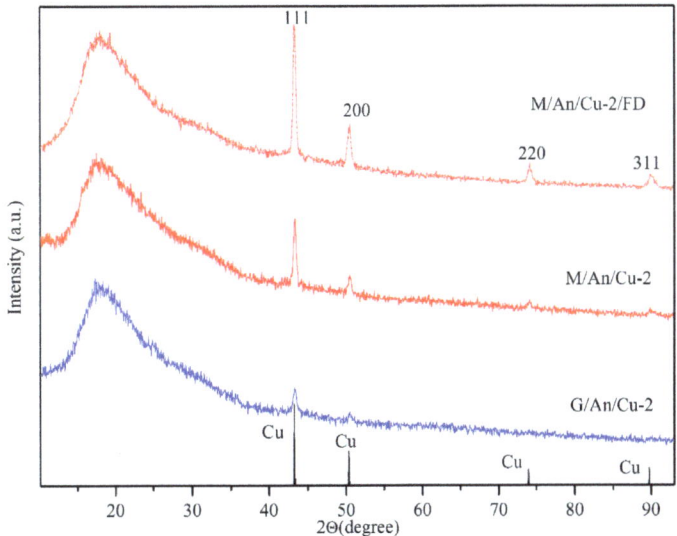

Figure 7. XRD patterns of reduction products M/An/Cu-2 and G/An/Cu-2 obtained by Method 2.

As might be expected, Method 2 in the case of macroreticular exchangers made it possible to prevent grain breakage. The M/An/Cu-2 (2 signifies Method 2) grains are undisturbed (Figure 8), and on their outer surface characteristic white spots are visible, identified as copper-rich areas (Figure 8a–e). The magnification of these areas reveals particles with a size and shape reminiscent of those previously identified on the surface of the G/An/Cu-1 (1 signifies Method 1) beads (Figure 8d). There are no characteristic bright objects inside the beads that would indicate the presence of copper-containing particles. The EDX analysis reveals that copper is dispersed equally in their entire volume (Figure 8h). Figure 9 presents a microscopic analysis of the surface and interior of the G/An/Cu-2 exchanger. In contrast to grains of the same anion exchangers obtained by Method 1 (G/An/Cu-1), grains obtained by a modified Method 2 do not show extensive cracks on the outer surface. Only some beads exhibit, as in Method 1, a mesh of small cracks on the surface (Figure 9a,b). An enlarged image of these cracked areas reveals numerous copper particles of the same shape and size as in the case of Method 1 (Figure 9c–e). As in the case of G/An/Cu-1, numerous copper aggregates with a size of about 1 µm are visible on the cross-section of G/An/Cu-2 grains (Figure 9f–h). However, in this case, the aggregates are formed of loosely packed particles with a size of about 200 nm. Evidently, because in the first method the G/An/Cu$_2$O samples were dried before the next reduction step, the matrix swelling and then its collapse caused more packing of the resulting copper deposit in G/An/Cu-1. Skipping the drying stage resulted in the lack of shrinkage and a much less packed distribution of the resulting copper particles in the G/An/Cu-2 beads (Figure 9g,h).

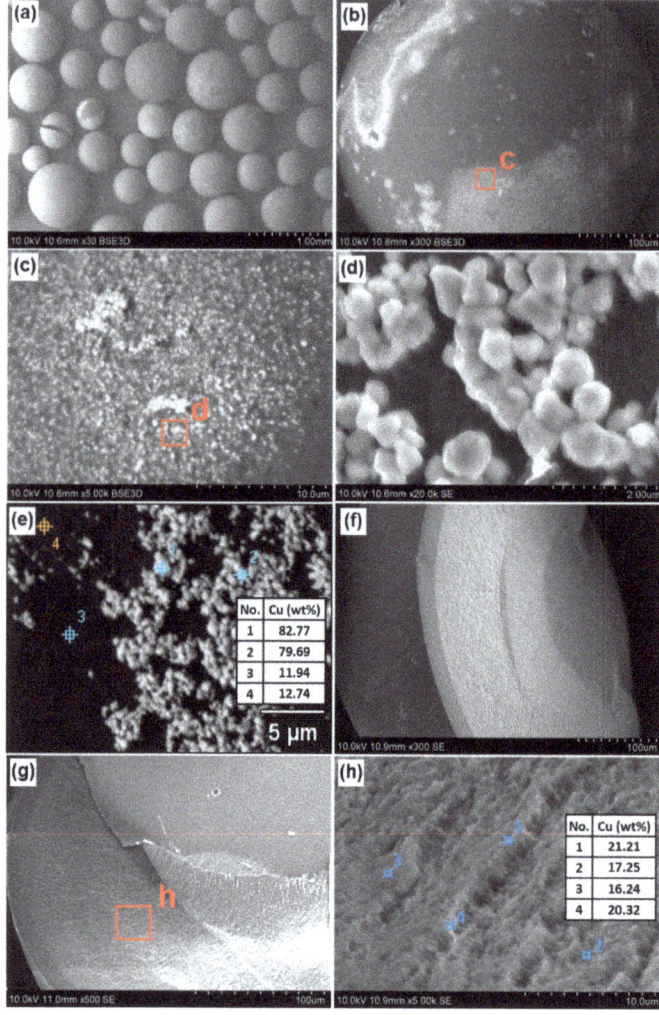

Figure 8. SEM images and EDS analysis of the M/An/Cu-2 (**a–e**) surface, (**f–h**) inside.

To avoid significant deformation and eliminate the shrinkage or toughening of the macroreticular M/An/Cu HIX beads resulting from the drying stage, the samples obtained by Method 2, instead of being conventionally dried after the reduction, were also freeze dried. Figure 10 shows photographs of the surface and interior of M/An/Cu-2 beads after freeze drying. On the surface of the macroreticular exchanger beads after lyophilization, numerous white areas can be observed, the magnification of which reveals numerous fine copper particles (Figure 10a–c). These copper particles appear to be more separated after freeze drying (Figure 10d,e) than what is observed for conventionally dried M/An/Cu-2 samples. Evenly dispersed white objects are visible on the cross-section images of freeze-dried grains (Figure 10f–h). These asymmetrical, star-like objects contain copper, and their size can be approximately estimated to be 0.2–1 μm (Figure 10h).

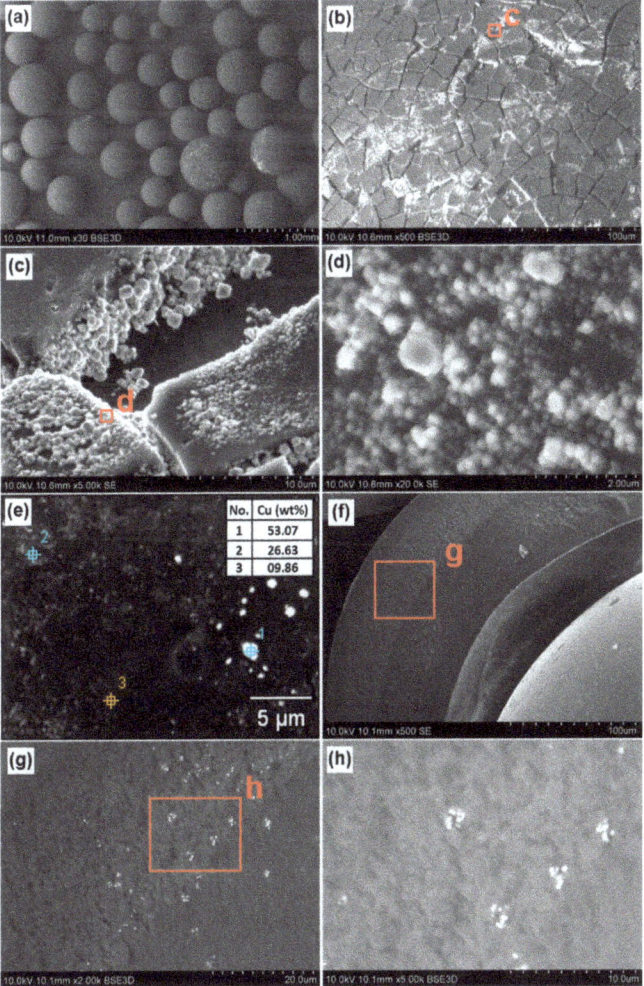

Figure 9. SEM images and EDS analysis of the G/An/Cu-2 (**a–e**) surface, (**f–h**) inside.

On the cross-section images of the conventionally dried M/An/Cu-2 beads, copper particles were not visible. Clearly, the M/An/Cu-2 drying method affects the shape and size of the deposit. Lyophilization involves two stages, namely freezing, in which freezing water forms ice crystals, and drying, when a vacuum is applied to sublimate the ice. Recent studies on nanoparticle lyophilization have shown that this method of drying causes different stresses during both stages of the process, contributing to an increased aggregation and agglomeration of nanoparticles [51,52]. Apparently, the growth of ice crystals in the pores of the M/An/Cu-2 macroreticular exchanger affected the deposit structure, resulting in the formation of large crystalline copper particles. The crystallization of ice caused mechanical stress on copper deposited in pores, leading not only to a denser packing of the deposit but also to an increase in the degree of its crystallization. A higher content of crystalline copper in freeze-dried samples was confirmed by XRD measurements (see the M/An/Cu-2/FD line in Figure 7). The analysis of porous characteristics of M/An/Cu-2 showed slightly lower porosity in the region of the macropores and also a very narrow pore size distribution in the range of the mesopores in comparison with the previous sample (M/An/Cu/FD) (Table 5, Figure 5c,f). The implication of this result is that elimination of operations conducted under elevated temperatures during Cu^0 deposition preserves the mechanical resistance and uniformity of porous characteristics of the host anion exchanger [48]. Simultaneously, larger crystalline Cu particles formed during lyophilization caused a slight decrease in BET surface area and an insignificant increase in mean pore diameter.

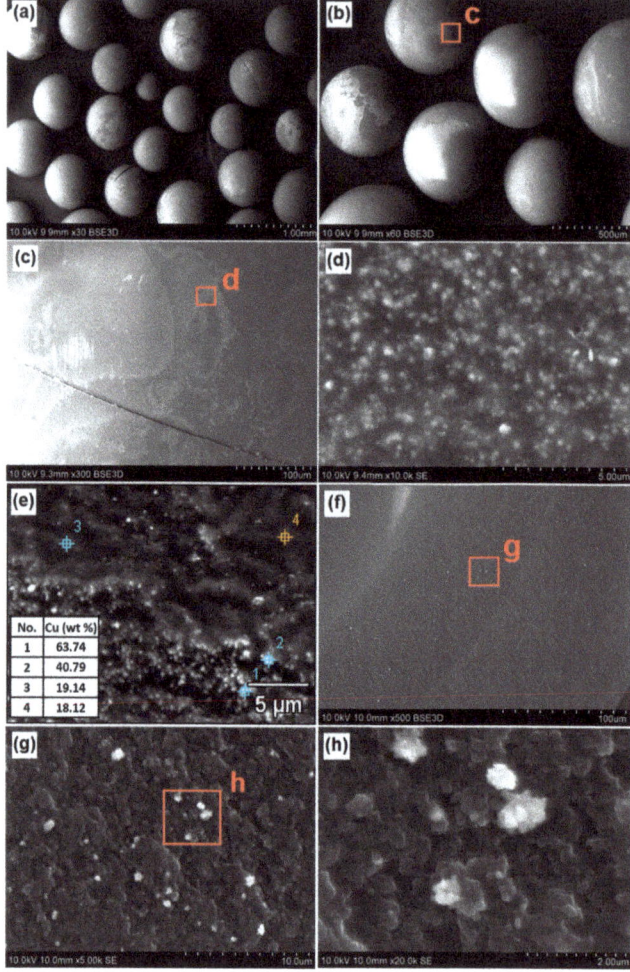

Figure 10. SEM images and EDS analysis of the M/An/Cu-2/FD (**a–e**) surface, (**f–h**) inside.

4. Conclusions

Fine Cu_2O particles dispersed in the anion exchanger phase can be easily transformed into Cu^0 particles, which after the reaction remain in the anion exchanger phase. The transformation proceeds fairly quantitatively in the matrix of both the macroreticular anion exchanger and the gel-like anion exchanger. Ascorbic acid (an ecofriendly reducing agent) is the reagent that makes this transformation possible. The reduction also proceeds in a sodium ascorbate solution but at a slower rate, despite the fact that the redox potential of this solution is lower than that of the ascorbic acid solution. A major benefit resulting from the use of ascorbic acid is that during reduction, the anion exchanger's functional groups are transformed into the ascorbate form, thereby protecting the metallic copper particle against oxidation (ascorbic acid is a well-known antioxidant). Having such substrates as preformed M/An/Cu$_2$O and G/An/Cu$_2$O (rich in the copper precursor) as templates for reductive conversion of Cu$_2$O to Cu^0, we managed in mild conditions to obtain HIXs with 7.0 and 5.0 wt% of the total Cu content.

While the conditions of the reduction reaction influenced the kinetics and efficiency of the process, as shown by the SEM EDS analyses, the morphology and distribution of copper within the structure of polymeric beads were mainly determined by the type of porosity of the supporting polymer and the distribution of preformed Cu$_2$O particles in the beads of the starting material. The G/An/Cu matrix contained loosely connected copper particles that formed clusters about 1 µm in size, whereas M/An/Cu beads contained clusters of closely lying individual particles with a size up to a maximum of 200 nm. In turn, the porous structure and integrity of the beads were dependent on the procedure of the whole process of the drying operations before and after the reduction reaction. The best products were obtained by using wet M/An/Cu$_2$O or G/An/Cu$_2$O as a starting material, conducting the reduction over 30 min at 50 °C, under acidic pH and freeze drying the final product. This procedure enabled us to obtain material with uniformly distributed star-like particles of ZVC with a high degree of crystallization and simultaneously with a preserved porous structure of the virgin supporting anion exchanger. In contrast, thermally dried products may exhibit a higher resistance for oxidation due to collapsed porous structure resulting lower gas permeability. Considering that both reaction parameters and operational techniques influence the size, shape, and distribution of Cu^0, as well as the porous structure of polymeric matrix, the obtained composite materials may have promising applications, including chemical catalytic or photocatalytic reactions involving both gaseous or liquid reagents as well as antibacterial activity.

Author Contributions: Conceptualization, E.K.-B.; methodology E.K.-B., I.J.-S.; investigation, I.J.-S., E.S., A.B., M.J.; resourses, E.S.; writing original draft, E.K.-B, I.J.-S., A.B., M.J.; writing-review and editing, I.J.-S.; visualization, I.J.-S., E.S., A.B., M.J.; Supervision, E.K.-B. All authors have read and agreed to the published version of the manuscript.

Funding: This research received no external funding.

Conflicts of Interest: The authors declare no conflict of interest.

References

1. Tamayo, L.; Azócar, M.; Kogan, M.; Riveros, A.; Páez, M. Copper-polymer nanocomposites: An excellent and cost-effective biocide for use on antibacterial surfaces. *Mater. Sci. Eng. C* **2016**, *69*, 1391–1409. [CrossRef]
2. Deka, P.; Borah, B.J.; Saikia, H.; Bharali, P. Cu-based nanoparticles as emerging environmental catalysts. *Chem. Rec.* **2019**, *19*, 462–473. [CrossRef]
3. Gawande, M.B.; Goswami, A.; Felpin, F.-X.; Asefa, T.; Huang, X.; Silva, R.; Zou, X.; Zboril, R.; Varma, R.S. Cu and Cu-based nanoparticles: Synthesis and applications in catalysis. *Chem. Rev.* **2016**, *116*, 3722–3811. [CrossRef]
4. Vincent, M.; Hartemann, P.; Engels-Deutsch, M. Antimicrobial applications of copper. *Int. J. Hyg. Environ. Health* **2016**, *219*, 585–591. [CrossRef] [PubMed]
5. Palza, H.; Nuñez, M.; Bastías, R.; Delgado, K. In situ antimicrobial behavior of materials with copper-based additives in a hospital environment. *Int. J. Antimicrob. Agents* **2018**, *51*, 912–917. [CrossRef]
6. Ojha, N.K.; Zyryanov, G.V.; Majee, A.; Charushin, V.N.; Chupakhin, O.N.; Santra, S. Copper nanoparticles as inexpensive and efficient catalyst: A valuable contribution in organic synthesis. *Coord. Chem. Rev.* **2017**, *353*, 1–57. [CrossRef]
7. Ali, Z.I.; Ghazy, O.A.; Meligi, G.; Saleh, H.H.; Bekhit, M. Copper nanoparticles: Synthesis, characterization and its application as catalyst for *p*-nitrophenol reduction. *J. Inorg. Organomet. Polym. Mater.* **2018**, *28*, 1195–1205. [CrossRef]
8. Sharma, P.; Pant, S.; Poonia, P.; Kumari, S.; Dave, V.; Sharma, S. Green synthesis of colloidal copper nanoparticles capped with Tinospora cordifolia and its application in catalytic degradation in textile dye: An ecologically sound approach. *J. Inorg. Organomet. Polym. Mater.* **2018**, *28*, 2463–2472. [CrossRef]

9. Rehman, M.; Liu, L.; Wang, Q.; Saleem, M.H.; Bashir, S.; Ullah, S.; Peng, D. Copper environmental toxicology, recent advances, and future outlook: A review. *Environ. Sci. Pollut. Res.* **2019**, *26*, 18003–18016. [CrossRef] [PubMed]
10. Zhao, X.; Wu, Y.; Xing, D.; Ren, Z.; Ye, L. Enhanced abatement of organic contaminants by zero-valent copper and sulfite. *Environ. Chem. Lett.* **2020**, *18*, 237–241. [CrossRef]
11. Mott, D.; Galkowski, J.; Wang, L.; Luo, J.; Zhong, C.-J. Synthesis of size-controlled and shaped copper nanoparticles. *Langmuir* **2007**, *23*, 5740–5745. [CrossRef] [PubMed]
12. Biçer, M.; Şişman, İ. Controlled synthesis of copper nano/microstructures using ascorbic acid in aqueous CTAB solution. *Powder Technol.* **2010**, *198*, 279–284. [CrossRef]
13. AL-Thabaiti, S.A.; Obaid, A.Y.; Khan, Z.; Bashir, O.; Hussain, S. Cu nanoparticles: Synthesis, crystallographic characterization, and stability. *Colloid Polym. Sci.* **2015**, *293*, 2543–2554. [CrossRef]
14. Dang, T.M.D.; Le, T.T.T.; Fribourg-Blanc, E.; Dang, M.C. The influence of solvents and surfactants on the preparation of copper nanoparticles by a chemical reduction method. *Adv. Nat. Sci. Nanosci. Nanotechnol.* **2011**, *2*. [CrossRef]
15. Andal, V.; Buvaneswari, G. Effect of reducing agents in the conversion of Cu_2O nanocolloid to Cu nanocolloid. *Eng. Sci. Technol. Int. J.* **2017**, *20*, 340–344. [CrossRef]
16. Xiong, J.; Wang, Y.; Xue, Q.; Wu, X. Synthesis of highly stable dispersions of nanosized copper particles using L-ascorbic acid. *Green Chem.* **2011**, *13*, 900–904. [CrossRef]
17. Jardón-Maximino, N.; Pérez-Alvarez, M.; Sierra-Ávila, R.; Ávila-Orta, C.A.; Jiménez-Regalado, E.; Bello, A.M.; González-Morones, P.; Cadenas-Pliego, G. Oxidation of copper nanoparticles protected with different coatings and stored under ambient conditions. *J. Nanomater.* **2018**, *2018*, 9512768. [CrossRef]
18. Usman, M.S.; Ibrahim, N.A.; Shameli, K.; Zainuddin, N.; Yunus, W.M.Z.W. Copper nanoparticles mediated by chitosan: Synthesis and characterization via chemical methods. *Molecules* **2012**, *17*, 14928–14936. [CrossRef]
19. Biswas, P.; Bandyopadhyaya, R. Demonstration of complete bacterial-killing in water at a very high flow-rate: Use of a surface-enhanced copper nanoparticle with activated carbon as a hybrid. *J. Chem. Technol. Biotechnol.* **2018**, *93*, 508–517. [CrossRef]
20. Olad, A.; Alipour, M.; Nosrati, R. The use of biodegradable polymers for the stabilization of copper nanoparticles synthesized by chemical reduction method. *Bull. Mater. Sci.* **2017**, *40*, 1013–1020. [CrossRef]
21. Ilnicka, A.; Walczyk, M.; Lukaszewicz, J.P.; Janczak, K.; Malinowski, R. Antimicrobial carbon materials incorporating copper nano-crystallites and their PLA composites. *J. Appl. Polym. Sci.* **2016**, *133*, 43429. [CrossRef]
22. Coneo Rodríguez, R.; Yate, L.; Coy, E.; Martínez-Villacorta, Á.M.; Bordoni, A.V.; Moya, S.; Angelomé, P.C. Copper nanoparticles synthesis in hybrid mesoporous thin films: Controlling oxidation state and catalytic performance through pore chemistry. *Appl. Surf. Sci.* **2019**, *471*, 862–868. [CrossRef]
23. Ríos, P.L.; Povea, P.; Cerda-Cavieres, C.; Arroyo, J.L.; Morales-Verdejo, C.; Abarca, G.; Camarada, M.B. Novel in situ synthesis of copper nanoparticles supported on reduced graphene oxide and its application as a new catalyst for the decomposition of composite solid propellants. *RSC Adv.* **2019**, *9*, 8480–8489. [CrossRef]
24. Sarkar, S.; Guibal, E.; Quignard, F.; SenGupta, A.K. Polymer-supported metals and metal oxide nanoparticles: Synthesis, characterization, and applications. *J. Nanoparticle Res.* **2012**, *14*, 715. [CrossRef]
25. Samiey, B.; Cheng, C.H.; Wu, J. Organic-inorganic hybrid polymers as adsorbents for removal of heavy metal ions from solutions: A review. *Materials* **2014**, *7*, 673–726. [CrossRef]
26. Smith, R.C.; Li, J.; Padungthon, S.; Sengupta, A.K. Nexus between polymer support and metal oxide nanoparticles in hybrid nanosorbent materials (HNMs) for sorption/desorption of target ligands. *Front. Environ. Sci. Eng.* **2015**, *9*, 929–938. [CrossRef]
27. Zhao, X.; Lv, L.; Pan, B.; Zhang, W.; Zhang, S.; Zhang, Q. Polymer-supported nanocomposites for environmental application: A review. *Chem. Eng. J.* **2011**, *170*, 381–394. [CrossRef]
28. Sarkar, S.; Chatterjee, P.K.; Cumbal, L.H.; SenGupta, A.K. Hybrid ion exchanger supported nanocomposites: Sorption and sensing for environmental applications. *Chem. Eng. J.* **2011**, *166*, 923–931. [CrossRef]
29. Simeonidis, K.; Mourdikoudis, S.; Kaprara, E.; Mitrakas, M.; Polavarapu, L. Inorganic engineered nanoparticles in drinking water treatment: A critical review. *Environ. Sci. Water Res. Technol.* **2016**, *2*, 43–70. [CrossRef]

30. Khalaj, M.; Kamali, M.; Khodaparast, Z.; Jahanshahi, A. Copper-based nanomaterials for environmental decontamination—An overview on technical and toxicological aspects. *Ecotoxicol. Environ. Saf.* **2018**, *148*, 813–824. [CrossRef]
31. Dzhardimalieva, G.I.; Uflyand, I.E. Preparation of metal-polymer nanocomposites by chemical reduction of metal ions: Functions of polymer matrices. *J. Polym. Res.* **2018**, *25*, 255. [CrossRef]
32. Kravchenko, T.A.; Chayka, M.Y.; Bulavina, E.V.; Glotov, A.V.; Yaroslavtsev, A.B. Formation of nanosized copper clusters in ion-exchange matrix. *Dokl. Phys. Chem.* **2010**, *433*, 111–114. [CrossRef]
33. Zolotukhina, E.V.; Kravchenko, T.A. Synthesis and kinetics of growth of metal nanoparticles inside ion-exchange polymers. *Electrochim. Acta* **2011**, *56*, 3597–3604. [CrossRef]
34. Kravchenko, T.A.; Sakardina, E.A.; Kalinichev, A.I.; Zolotukhina, E.V. Stabilization of copper nanoparticles with volume- and surface-distribution inside ion-exchange matrices. *Russ. J. Phys. Chem. A* **2015**, *89*, 1648–1654. [CrossRef]
35. Jacukowicz-Sobala, I.; Kociołek-Balawejder, E.; Stanisławska, E.; Dworniczek, E.; Seniuk, A. Antimicrobial activity of anion exchangers containing cupric compounds against *Enterococcus faecalis*. *Colloids Surfaces A Physicochem. Eng. Asp.* **2019**, *576*, 103–109. [CrossRef]
36. Kociołek-Balawejder, E.; Stanisławska, E.; Jacukowicz-Sobala, I.; Baszczuk, A.; Jasiorski, M. Deposition of spherical and bracelet-like Cu_2O nanoparticles within the matrix of anion exchangers via reduction of tetrachlorocuprate anions. *J. Environ. Chem. Eng.* **2020**, *8*, 103722. [CrossRef]
37. Umer, A.; Naveed, S.; Ramzan, N.; Rafique, M.S.; Imran, M. A green method for the synthesis of copper nanoparticles using l-ascorbic acid. *Rev. Mater.* **2014**, *19*, 197–203. [CrossRef]
38. Wu, S. Preparation of fine copper powder using ascorbic acid as reducing agent and its application in MLCC. *Mater. Lett.* **2007**, *61*, 1125–1129. [CrossRef]
39. Liu, Q.; Yasunami, T.; Kuruda, K.; Okido, M. Preparation of Cu nanoparticles with ascorbic acid by aqueous solution reduction method. *Trans. Nonferrous Met. Soc. China* **2012**, *22*, 2198–2203. [CrossRef]
40. Ramos, A.R.; Tapia, A.K.G.; Piñol, C.M.N.; Lantican, N.B.; del Mundo, M.L.F.; Manalo, R.D.; Herrera, M.U. Effects of reaction temperatures and reactant concentrations on the antimicrobial characteristics of copper precipitates synthesized using L-ascorbic acid as reducing agent. *J. Sci. Adv. Mater. Devices* **2019**, *4*, 66–71. [CrossRef]
41. Feng, Z.; Wei, W.; Wang, L.; Hong, R. Cross-linked PS-DVB/Fe_3O_4 microspheres with quaternary ammonium groups and application in removal of nitrate from water. *RSC Adv.* **2015**, *5*, 96911–96917. [CrossRef]
42. Shephard, A.B.; Nichols, S.C.; Braithwaite, A. Moisture induced solid phase degradation of L-ascorbic acid part 3, structural characterisation of the degradation products. *Talanta* **1999**, *48*, 607–622. [CrossRef]
43. Lee, Y.-J.; Kim, K.; Shin, I.-S.; Shin, K.S. Antioxidative metallic copper nanoparticles prepared by modified polyol method and their catalytic activities. *J. Nanoparticle Res.* **2020**, *22*, 8. [CrossRef]
44. Kawamura, G.; Alvarez, S.; Stewart, I.E.; Catenacci, M.; Chen, Z.; Ha, Y.C. Production of oxidation-resistant Cu-based nanoparticles by wire explosion. *Sci. Rep.* **2015**, *5*, 1–8. [CrossRef]
45. Chen, C.S.; Chen, C.C.; Lai, T.W.; Wu, J.H.; Chen, C.H.; Lee, J.F. Water adsorption and dissociation on Cu nanoparticles. *J. Phys. Chem. C* **2011**, *115*, 12891–12900. [CrossRef]
46. Traboulsi, A.; Dupuy, N.; Rebufa, C.; Sergent, M.; Labed, V. Investigation of gamma radiation effect on the anion exchange resin Amberlite IRA-400 in hydroxide form by Fourier transformed infrared and 13C nuclear magnetic resonance spectroscopies. *Anal. Chim. Acta* **2012**, *717*, 110–121. [CrossRef]
47. Ramanarayanan, T.A.; Alonzo, J. Oxidation of copper and reduction of Cu_2O in an environmental scanning electron microscope at 800 °C. *Oxid. Met.* **1985**, *24*, 17–27. [CrossRef]
48. Kociołek-Balawejder, E.; Stanisławska, E.; Ciechanowska, A. Iron(III) (hydr)oxide loaded anion exchange hybrid polymers obtained via tetrachloroferrate ionic form—Synthesis optimization and characterization. *J. Environ. Chem. Eng.* **2017**, *5*, 3354–3361. [CrossRef]
49. Dzyazko, Y.; Volfkovich, Y.; Perlova, O.; Ponomaryova, L.; Perlova, N.; Kolomiets, E. Effect of porosity on ion transport through polymers and polymer-based composites containing inorganic nanoparticles (Review). In *Nanophotonics, Nanooptics, Nanobiotechnology, and Their Applications*; Fesenko, O., Yatsenko, L., Eds.; Springer International Publishing: Cham, Switzerland, 2019; pp. 235–253.
50. Kociołek-Balawejder, E.; Stanisławska, E.; Mucha, I. Freeze dried and thermally dried anion exchanger doped with iron(III) (hydr)oxide—Thermogravimetric studies. *Thermochim. Acta* **2019**, *680*, 178359. [CrossRef]

51. Beirowski, B.J.; Gieseler, H. Stabilisation of nanoparticles during freeze drying: The difference to proteins. *Eur. Pharm. Rev.* **2011**, *31*, 1–11.
52. Abdelwahed, W.; Degobert, G.; Stainmesse, S.; Fessi, H. Freeze-drying of nanoparticles: Formulation, process and storage considerations. *Adv. Drug Deliv. Rev.* **2006**, *58*, 1688–1713. [CrossRef] [PubMed]

Publisher's Note: MDPI stays neutral with regard to jurisdictional claims in published maps and institutional affiliations.

 © 2020 by the authors. Licensee MDPI, Basel, Switzerland. This article is an open access article distributed under the terms and conditions of the Creative Commons Attribution (CC BY) license (http://creativecommons.org/licenses/by/4.0/).

Article

Polymersome Poration and Rupture Mediated by Plasmonic Nanoparticles in Response to Single-Pulse Irradiation

Gina M. DiSalvo [1,†], Abby R. Robinson [1,†], Mohamed S. Aly [2], Eric R. Hoglund [3], Sean M. O'Malley [2,4] and Julianne C. Griepenburg [2,4,*]

1. Department of Chemistry, Rutgers University-Camden, 315 Penn Street, Camden, NJ 08102, USA; gdisalvo22@gmail.com (G.M.D.); robinsonabby13@gmail.com (A.R.R.)
2. Department of Physics, Rutgers University-Camden, 227 Penn Street, Camden, NJ 08102, USA; msa194@scarletmail.rutgers.edu (M.S.A.); omallese@camden.rutgers.edu (S.M.O.)
3. Department of Materials Science and Engineering, University of Virginia, Thornton Hall, P.O. Box 400259, Charlottesville, VA 22904, USA; erh3cq@virginia.edu
4. Center for Computational and Integrative Biology, Rutgers University-Camden, Camden, NJ 08102, USA
* Correspondence: j.griepenburg@rutgers.edu; Tel.: +1-856-225-6293
† Denotes equal contribution.

Received: 26 September 2020; Accepted: 13 October 2020; Published: 16 October 2020

Abstract: The self-assembly of amphiphilic diblock copolymers into polymeric vesicles, commonly known as polymersomes, results in a versatile system for a variety of applications including drug delivery and microreactors. In this study, we show that the incorporation of hydrophobic plasmonic nanoparticles within the polymersome membrane facilitates light-stimulated release of vesicle encapsulants. This work seeks to achieve tunable, triggered release with non-invasive, spatiotemporal control using single-pulse irradiation. Gold nanoparticles (AuNPs) are incorporated as photosensitizers into the hydrophobic membrane of micron-scale polymersomes and the cargo release profile is controlled by varying the pulse energy and nanoparticle concentration. We have demonstrated the ability to achieve immediate vesicle rupture as well as vesicle poration resulting in temporal cargo diffusion. Additionally, changing the pulse duration, from femtosecond to nanosecond, provides mechanistic insight into the photothermal and photomechanical contributors that govern membrane disruption in this polymer–nanoparticle hybrid system.

Keywords: polymersomes; vesicles; nanoparticles; drug-delivery; ultrafast laser; plasmonic; nanobubble; fragmentation

---

1. Introduction

The ability to deliver cargo from carrier vesicles with high spatiotemporal control would greatly enhance a variety of applications, including nanoreactors, microreactors, and drug delivery systems. Liposomes have been an extensively investigated carrier system for targeted release and their implementation in nanomedicine has resulted in significant advances in therapeutic compound stability, tissue uptake, and biodistribution in vivo [1]. However, many challenges exist with liposome systems such as long-term stability and membrane permeability. To overcome these limitations, attention has been redirected to their fully synthetic polymeric analogs, known as "polymersomes" [2].

Polymersomes are a class of fully synthetic vesicles which self-assembly from a variety of amphiphilic diblock copolymers [3]. The self-assembly process results in two distinct regions: a bilayer member that contains a hydrophobic center, and an aqueous core which can hold hydrophilic encapsulants. The properties of the bilayer membrane are readily tunable by adjusting the molecular weights (MW) of the hydrophobic and hydrophilic blocks of the amphiphile. For example, a hyperthick bilayer membrane up to 50 nm, allowing for storage of hydrophobic encapsulants, can be achieved by using high MW amphiphiles [4]. This robust, dual-compartment encapsulation ability makes polymersomes an ideal candidate for applications which require molecular transport in an unreacted state. For these systems to garner success in application, the ability to release cargo with high spatiotemporal control is of utmost importance. The design of stimuli responsive polymersomes has been extensively investigated, as summarized by Che et al., resulting in the development of polymersomes that are responsive to a variety of external stimuli (e.g., ultrasound, magnetic fields, light) or internal stimuli (e.g., temperature, pH, redox) [5].

Light is a particularly attractive trigger as it is minimally damaging yet deeply penetrating at certain wavelengths (and intensities). It can also be localized in time and space, and externally triggered without the need for additional reagents [5,6]. While there exist several examples of light stimulated encapsulant release from polymersomes, many require the use of high-energy (UV) light to trigger cleavable groups such as nitrobenzyl or azobenzene [7–10]. Examples of polymersomes that rupture in response to light in the visible to near-infrared (NIR) window involve the use of conjugated multi-porphyrin dyes encapsulated within the hydrophobic membrane [11]. Kamat et al. and Griepenburg et al. show vesicle deformation and rupture of micron- and nano-scale vesicles, respectively; however, this is achieved with long irradiation times using combined, continuous wave (CW) light at multiple wavelengths across the visible region [12–14].

Plasmonic nanoparticles are ideal candidates as photosensitizers to initiate light-triggered release due to the photothermal response accompanying excitation of their localized surface plasmon resonance (LSPR) and large optical cross sections. One particularly notable example includes the work by Amstad et al. where hydrophobic gold nanoparticles (NPs) (AuNPs) were incorporated into the membrane of polymersomes self-assembled from poly(ethylene glycol)-b-poly(lactic acid) (PEG-b-PLA) and poly(N-isopropylacrylamide)-b-poly(lactic-co-glycolic acid) (PNIPAM-b-PLGA) diblock copolymers [15]. Vesicles were irradiated with three continuous wave (CW) lasers simultaneously (488, 532, and 633 nm) resulting in a decrease in intact vesicles. Under irradiation, the thermoresponsive PNIPAM blocks underwent amphiphilicity changes in response to local hot spots generated by the AuNPs, ultimately resulting in complete destabilization of the bilayer. Additional work by Zhang et al. includes the use of porous silicon nanoparticles conjugated with gold nanorods that are encapsulated within the aqueous core of hybrid poly(ethylene glycol)-b-poly(lactic acid) nanopolymersomes to achieve hydrophobic and hydrophilic drug loading capacities as well as photothermal responsiveness. These nanovesicles were shown to suppress breast tumors in mice by up to 94% through loading with a triple-drug combination [16].

One drawback common to the aforementioned systems is the need for long irradiation times, which is inherent to the use of CW irradiation. Such long irradiation times are not ideal in dynamic systems as this can result in decreased localization of the released encapsulant. In contrast, pulsed, and in particular, ultrafast laser irradiation, holds great promise for the use in delivery applications, due primarily to the ability of achieving high peak power in short durations, thus allowing for a photomechanical response which would enhance localization.

Several notable examples exist for using nanoparticle excitation to trigger destabilization of liposomes and subsequent cargo release; however, only a few utilize pulsed laser systems [17,18]. In a closely related example, Mathiyazhakan et al. highlight a system where hydrophilic AuNPs are loaded into the core of a thermo-responsive liposome [19]. The ability to release a model drug, calcein, was demonstrated with pulsed laser irradiation, and the mechanism through which membrane disruption occurred was attributed to microbubble cavitation. This example, however, required several nanosecond pulses (25–100) to achieve a significant percentage of release. The authors acknowledge that CW irradiation failed to trigger release from their system. Wu et al. studied cargo release from liposomes with hollow gold nanoshells incorporated at various locations (tethered, freely outside, inside the aqueous core) under ultrafast irradiation, and determined that membrane–NP proximity is an important factor in release profiles [20]. These examples reinforce the benefits of polymersomes, where nanoparticles can more readily be incorporated into their relatively thick bilayer membrane, facilitating closer proximity, and thus more efficient release without the need for thermoresponsive diblock copolymers.

Herein, as shown in Figure 1, we demonstrate the use of 2.5 nm, hydrophobic, plasmonic gold nanoparticles (AuNP) incorporated into the hydrophobic membrane of micron-scale polymersomes as efficient photosensitizers for polymersome poration and/or rupture. The use of pulsed lasers is chosen to decrease irradiation time, and thereby potentially increase temporal resolution over conventional CW systems. The diblock copolymer poly(butadiene)-b-poly(ethylene oxide) PBD_{35}-b-PEO_{20} was chosen given that it is robust and not inherently photoresponsive or thermoresponsive in temperature ranges of interest for this study as well as future applications in vitro and in vivo [21]. Single-vesicle imaging studies demonstrate that encapsulant release from polymersomes can be effectively controlled via pulse energy and AuNP concentration. Additionally, the mechanisms by which membrane disruptions are brought about are investigated through laser pulse duration studies.

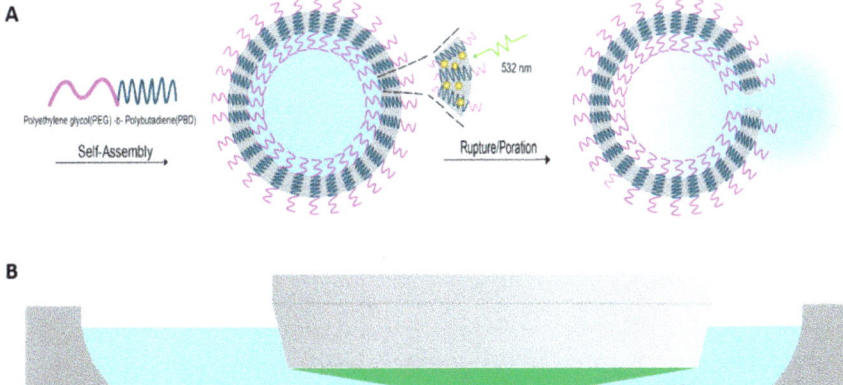

Figure 1. (**A**) Schematic representation of hydrophobic, 2.5 nm, plasmonic gold nanoparticles (AuNPs) incorporated into the hydrophobic membrane of micron-scale polyethylene oxide-*b*-polybutadiene (PBD_{35}-*b*-PEO_{20}) polymersomes during the self-assembly process. Single-pulse irradiation on resonance with the localized surface plasmon resonance (LSPR) of the AuNPs results in polymersome rupture or poration. (**B**) Irradiation studies are performed at the single-vesicle level, by focusing the laser through an optical objective which allows for irradiation and subsequent imaging.

2. Materials and Methods

2.1. Micron-Scale Polymersome Preparation

Polymersomes were prepared from the diblock copolymer, polyethylene oxide-*b*-polybutadiene (PBD_{35}-*b*-PEO_{20}) (Polymer Source, Montreal, QC, Canada) by the gel-assisted rehydration method [22]. The diblock copolymer was dissolved in chloroform at a concentration of 5 mg/mL. To visualize the vesicles under fluorescence microscopy, Nile red (Santa Cruz Biotechnology, Dallas, TX, USA) was added to the organic copolymer solution to yield a final concentration of 0.5 mol%. This step also includes the addition of 2–5 nm dodecanethiol functionalized spherical AuNPs (Alfa Aesar, Haverhill, MA, USA) to the organic solution, in concentrations varying from 0.013–0.066% (*w/v*). The organic solution was spread atop an agarose coated coverslip. The agarose coated coverslips were prepared by spreading a

1% (*w/v*) agarose solution (Sigma Aldrich, St. Louis, MO, USA) atop a 25 × 50 × 0.13 – 0.17 mm glass coverslip (Electron Microscopy Sciences, Hatfield, PA, USA) using a 1000 µL pipette tip. The agarose coated coverslips were then dried at 37 °C in a vacuum oven to fully remove the solvent, resulting in a thin agarose film. Organic solution (55 µL) containing copolymer, nanoparticle, and Nile red fluorophore was spread on the agarose coated coverslip with the side of a 12-gauge needle until the film visibly dried. The polymer coated coverslip was then further dried under vacuum for at least 1 h. A custom-made polydimethylsiloxane (PDMS) well, prepared from a Dow SYLGARDTM 170 Silicone Encapsulant Kit (Ellsworth Adhesives, Germantown, WI, USA), was adhered to the top of the polymer and agarose coated coverslip. This allows for the rehydration buffer to be added and contained on top of the polymer coated agarose coverslip. The rehydration buffer (600 µL) was added to the well. At this step, any hydrophilic components to be encapsulated within the polymersome core were added to the rehydration buffer. The hydrophilic fluorophore, 3–5k molecular weight fluorescein isothiocyanate-dextran (FITC-dextran) (Sigma Aldrich, St. Louis, MO, USA) was added to the 280 mM sucrose (Sigma Aldrich, MO, USA) rehydration buffer to a final concentration of 0.5 mg/mL. Upon addition of the rehydration buffer and heating ranging from 45 °C–65 °C for 45 min, polymersomes form and remain partially attached to the agarose gel. Figure S1 shows a schematic representation of the gel rehydration procedure. Vesicles prepared via this protocol had a mean diameter of 30.3 µm with a standard deviation of 8.5 µm. During the irradiation experiments, vesicles remain attached to the coverslip with the hydration well in place, to ensure that vesicles are static and in an aqueous environment.

2.2. Polymersome Imaging

Upon polymersome formation, the sample was imaged by fluorescence microscopy using a ZEISS Axio Examiner, fixed stage upright microscope, equipped with a 20× (0.5 numerical aperture) dipping objective which aided in visualization of the vesicles located at the bottom of the rehydration well, as depicted in Figure 1B. During the irradiation step, the fluorescent cube was rotated out of position to allow the beam to pass through to the objective unobstructed, and then moved back into position to allow for post-irradiation monitoring of the vesicle and its fluorescent encapsulants.

2.3. Removal of Excess Encapsulant

For polymersomes prepared with encapsulants, e.g., fluorescein isothiocyanate-dextran (FITC-dextran), within the vesicle core, a buffer exchange was necessary to rid the sample of the non-encapsulated components. This helped reduce the background fluorescence and facilitated imaging of the vesicle core. Two pieces of tubing (0.012 "IDx0.030" OD, Cole Parmer, Vernon Hills, IL, USA) were inserted into the buffer solution through the side walls of the custom PDMS well. Fresh buffer was slowly pumped into the well from one syringe pump, while the sample buffer containing free FITC-dextran was simultaneously pulled from the well using a second syringe pump (Figure 2). A gentle flow prevents vesicles from lifting off of the agarose gel.

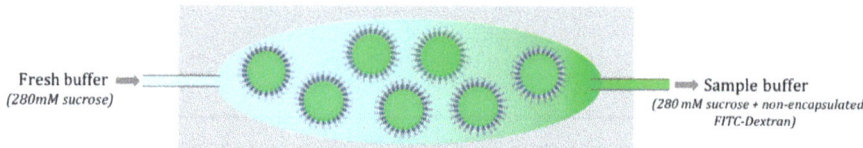

Figure 2. Top-down schematic for buffer exchange procedure. Buffer exchange is required for removal of non-encapsulated aqueous molecules. Fluorescein isothiocyanate-dextran (FITC-dextran) is included in the hydration buffer for encapsulation within the polymersome core. Excess FITC-dextran which remains in the buffer must be removed in order to facilitate imaging of the polymersome core.

2.4. Pulsed Laser Irradiation

Irradiation experiments were made feasible through a unique multi-functional optical microscope integrated with femtosecond and nanosecond laser systems. The ultrafast system is a Spectra Physics Solstice Ti:sapphire (100 fs) pumping a Light Continuum optical parametric amplifier (OPA) with the output set to 532 nm. The nanosecond laser is a Ekspla NL201 (7 ns) diode pumped Nd:YAG system equipped with a second harmonic generation crystal to an output wavelength of 532 nm. The beams are passed through a pair of filter wheels, to adjust attenuation, before being steered into an auxiliary port located above the filter carousel of the microscope and made collinear with the optical path via a dichroic mirror. This configuration, as demonstrated in Figure 1B and pictured in Figure S2, allows for irradiation and subsequent imaging of micron-sized vesicles. For micron-sized polymersomes, individual vesicles were singled-out and irradiated with a single pulse and subsequently imaged through a time-series to monitor vesicle integrity and encapsulant release. All imaging and irradiation experiments were carried out at 21 °C.

The laser spot size was measured and adjusted through the 20× dipping objective to ensure the laser beam was focused directly on the polymersome to be irradiated. This spot size was measured with a damage spot on a glass bottom petri dish where applied ink spots were submerged in water and irradiated as shown in Figure S3. The pulse energy was attenuated and measured using an energy meter placed on the microscope stage at the location of sample irradiation. No objective was used for this measurement, however, an associated 8% reduction in energy was taken into account when reporting pulse energy.

Zeiss ZenPro 2.5 software was utilized for analysis of the polymersome time series following irradiation. To quantify release, a region of interest (ROI) was selected inside of the polymersome as well as a background region to normalize the signal and account for photobleaching. A sample data series is shown in Table S1.

2.5. Studying the Fate of the Incorporated AuNPs upon Polymersome Irradiation

Polymersome samples containing 0.05% (*w/v*) AuNPs were prepared as described. Nano-polymersomes were self-assembled as described below and the sample was pipetted into a PDMS well placed atop a glass slide, to which a small magnetic stir bar was added. The sample was then placed under the dipping objective of the upright microscope and continuously irradiated at 532 nm from either the femtosecond system (80 nJ, 100 fs, 500 Hz) for 20 min or the nanosecond laser (10.3 µJ, 7 ns, 500 Hz) for 20 min. Gold nanoparticle size distributions were determined by high-resolution transmission electron microscopy using a 300 kV FEI Titan system equipped with a Gatan 794 camera, achieving a maximum resolution of 0.205 nm. Nanoparticle size distributions were determined by measuring the feret diameter for a minimum of 5000 particles across multiple images per sample using ImageJ. Representative images have been included in Figure S4.

2.6. Nano-Scale Polymersome Preparation

Nano-scale vesicles were prepared via the direct solvent injection method. The diblock copolymer, PEO_{20}-*b*-PBD_{35} (Polymer Source, Montreal, QC, Canada), was dissolved in tetrahydrofuran (THF) (Sigma Aldrich, St. Louis, MO, USA) to yield a concentration of 4 mg/mL, unless otherwise stated. Incorporation of various concentrations of AuNPs to the polymer-THF solution took place by a resuspension process where 2–5 nm dodecanethiol functionalized, spherical AuNPs (Alpha Aesar, Haverhill, MA, USA) suspended in toluene (2% *v/v*), were transferred to an open Eppendorf tube and gently heated to aid in solvent evaporation. The dry AuNPs were then re-suspended by adding 300 µL of the polymer–THF solution. To prepare for solvent injection, a stir bar was added to a glass vial containing 700 µL double deionized (DDI) (Milli-Q) water, and parafilm used to seal the vial. The vial containing the aqueous solution was subjected to magnetic stirring. The copolymer–AuNP–THF solution was drawn into a 3 mL syringe and a 23 gauge 1" blunt tip needle with tubing (0.012 "IDx0.030" OD) (Cole Parmer, Vernon Hills, IL, USA) was attached. The organic solution was added dropwise (10 µL/second) to the aqueous solution under continuous stirring, which created an emulsion resulting in the self-assembly of nano-polymersomes. The polymersome sample was then filtered using a 0.45 µm polytetrafluoroethylene (PTFE) syringe filter (GS-TEK, Newark, DE, USA) to remove any large aggregates which possibly formed from copolymer and/or AuNPs which did not assemble into vesicular structures.

2.7. Size Confirmation

Nano-polymersome size was determined by measuring the hydrodynamic diameter via dynamic light scattering (DLS) using a Malvern Zetasizer Nano ZS with a scattering angle set to 173°. The sample was filtered using a 0.45 µm syringe filter, as described above, to ensure that any large particulates (i.e., non-vesicular aggregates, dust) did not bias the measurement. The polymersome sample was diluted (~33-fold), using DDI H_2O, in a polystyrene cuvette to a final volume of 1 mL and the sample was measured in triplicate, with each run consisting of a minimum of 10 scans at 25 °C. While cryogenic transmission electron microscopy (cryo-TEM) is often regarded as the gold-standard for size determination of nano-vesicles smaller than 200 nm, DLS can be used as a simple screening technique when accompanied with careful analysis [23,24]. Nano-polymersome formation was screened by confirming that both the intensity distribution and Z-Average size were between 50 and 100 nm and the polydispersity index (PDI) was used to determine uniformity for each sample. Non-irradiated, nanosecond (ns)-irradiated, and femtosecond (fs)-irradiated samples were determined to have average hydrodynamic diameters by intensity of 71 nm (98.8%), 67 nm (98.5%), and 75 nm (98.2%), respectively, all with a PDI of 0.200 or less. As these samples were relatively monodisperse, the Z-Average sizes, based on the harmonic mean calculated within the software (Zetasizer v7.11), correlate well with intensity distributions, at 61.2 nm, 60.4 nm, and 63.0 nm.

3. Results

3.1. AuNPs as Photosensitizers for Complete Polymersome Rupture

Initial studies sought to determine the sensitivity of micron-scale PEO_{20}-*b*-PBD_{35} polymersomes to single-pulse, femtosecond laser irradiation, and whether the addition of AuNPs within the hydrophobic membrane would increase photosensitivity.

As shown in Figure 3, complete vesicle rupture is observed immediately following irradiation with a single pulse at 532 nm, while the surrounding, non-irradiated vesicles remain intact. As a control, polymersomes without AuNPs in the hydrophobic membrane were subjected to identical fs irradiation and it was determined through interpolation that 5× the energy was required to initiate rupture. It follows that complete vesicle rupture would result in immediate content release of hydrophilic encapsulants to the surrounding environment.

To further investigate the effect of AuNP encapsulation within the hydrophobic membrane of the polymersome, Figure 3C shows the rupture frequency as a function of pulse energy for both populations of vesicles with and without AuNPs. No fewer than 20 similarly sized vesicles were irradiated at each energy and only vesicles with no more than one contact point with another vesicle were chosen for this study. Vesicles selected for irradiation had an average diameter of 45.2 ± 6.8 µm. As shown in Figure 3C, pulse energy required to induce observable rupture for vesicles containing AuNPs with fs-irradiation was 0.21 µJ, thus indicating a rupture threshold in the 0.057–0.21 µJ range. The energy required to observe rupture in vesicles w/o AuNPs is interpolated to be 5x higher at 0.73 µJ. Thus, vesicles with AuNPs present in the membrane displayed significantly greater photosensitivity under fs irradiation.

Under ns-irradiation, the difference in energy to bring about similar rupture percentages was even greater than the case of fs-irradiation, as shown in Figure 3D. Vesicles without AuNPs experienced a rupture rate of 100% at 15 µJ, which decreases to 39% at 9.3 µJ, and by 3.7 µJ, rupture was no longer observed. Vesicles with AuNPs show a distinct reduction in energy to bring about rupture. Even at the lowest pulse energy of 0.37 µJ, polymersomes containing plasmonic nanoparticles still experience rupture at a frequency of 17%. This represents an interpolated 14-fold decrease in the rupture threshold for polymersomes containing AuNPs under ns-irradiation. Moreover, at 9.3 µJ, vesicles with AuNPs ruptured 100% of the time in comparison to only 38% for those without.

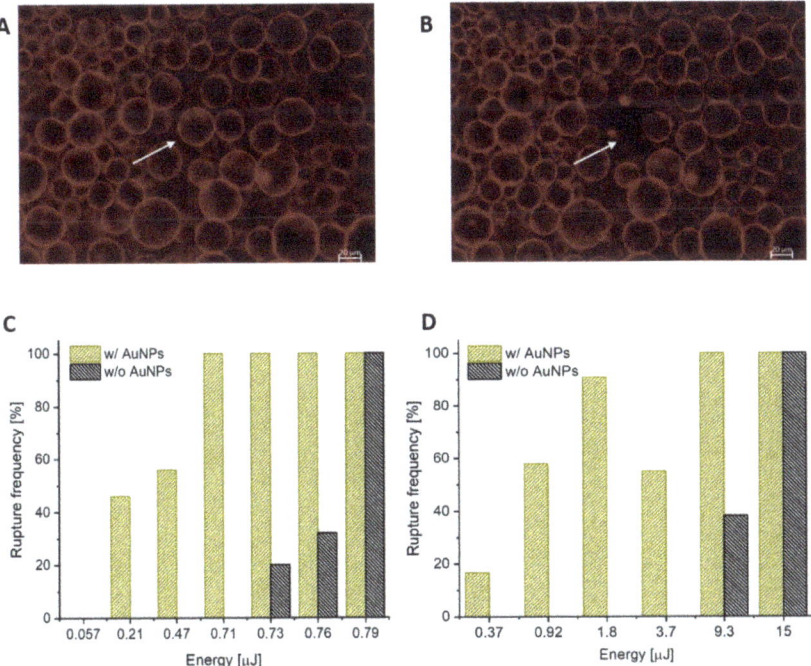

Figure 3. (**A**) Polymersomes with Nile red and 0.066% (*w/v*) AuNPs encapsulated in the hydrophobic region of the membrane, prior to laser irradiation. (**B**) The polymersome indicated by the white arrow was irradiated with a single 532 nm fs pulse and immediately imaged. Complete rupture of the polymersome is observed while the surrounding vesicles remain intact. Scale bar represents 20 µm. (**C**) Rupture statistics for polymersomes upon femtosecond irradiation. Rupture frequency of polymersomes with AuNPs within the membrane (gold) and without AuNPs (black) are shown. These percentages represent the number of vesicles undergoing complete rupture upon irradiation by a single 532 nm fs pulse at the specified energies. A minimum of 20 vesicles were irradiated per pulse energy. (**D**) Rupture statistics for polymersomes using nanosecond irradiation. The percentage of polymersomes rupturing in response to varying ns pulse energies was determined for populations of vesicles with (gold) and without (black) AuNPs in the membrane. These percentages represent complete rupture upon irradiation by a single 532 nm, ns pulse at the specified energies.

3.2. Using Pulse Duration for Mechanistic Insight

The results above clearly suggest enhanced photosensitivity of vesicles containing plasmonic nanoparticles in the membrane in response to both fs and ns pulses. However, the vast difference in required pulse energies prompts questioning from a mechanistic standpoint given the differences in pulse durations, i.e., 100 fs vs. 7 ns. After excitation of the AuNPs within the membrane, energy must escape the excited AuNPs; the mechanism by which this occurs will have an effect on the self-assembly of the copolymer and ultimately determine the threshold for rupture. Primarily thermal mechanisms will allow for heat to diffuse away from the particle and into the surrounding copolymer, thus altering self-assembly. Alternatively, more energetic effects such as the formation of a rapidly expanding vapor bubble around the particle, or fragmentation of the AuNP by coulombic explosion can occur, leading to membrane disruption. The reduction in energy between fs and ns plasmonic excitation for 2.5 nm AuNPs can be explained, at least from a thermal standpoint, by a recent article published by Metwally et al. [25]. In their article, they investigated the fluence required for the water immediately outside of variously-sized AuNPs to reach the spinodal point which initiates formation of a vapor layer around the particle. They looked at both ns and fs pulse durations with a wavelength 532 nm. Results showed that, for sub-20 nm AuNPs, the fluence needed to reach the spinodal temperature was significantly larger for nanosecond excitation than with femtosecond irradiation. Thus, it is proposed that the thermal dissipation of energy out of the particles during delivery of a ns pulse suppresses the peak temperature of the water layer. Femtosecond irradiation does not experience this thermal-dissipation during the pulse duration because all of the incident energy is delivered to the particle in a time frame less than the electron–phonon (~1 ps) and phonon–phonon (~100 ps) relaxation times [26]. Hence, the corresponding thermal energy is delivered essentially at once to the water surrounding the particle and a higher temperature can be attained for a given fluence. In the case of AuNPs loaded in the membrane of polymersomes, reaching the spinodal temperature of water may not be critical, given that PBD has a relatively low water permeability, however it is reasonable to conclude that some critical temperature exists that initiates membrane disruption such that rupture results, especially in the case of ns irradiation. Thus, it is expected that nanoparticle energy dissipation will be attributed to different primary processes within the two regimes.

To further probe the mechanistic differences in rupture initiation, the fate of the nanoparticles in the membrane both before and after ns and fs irradiation were investigated using high-resolution transmission electron microscopy (HR-TEM). The pulse durations used, 100 fs and 7 ns, are of particular interest as they lie on either side of the electron phonon coupling time [27]. While micron-scale vesicles provide an excellent size regime to facilitate single vesicle imaging, polymersomes on the nano-scale are better suited to examine the fate of the nanoparticles in response to excitation because all loaded particles within a vesicle will be subject to irradiation due to the smaller size of the vesicle in comparison to the beam profile. Stirring of the colloidal solution ensures exposure of the entire population of vesicles during continuous pulsed irradiation.

The AuNP size distribution along with the mode for each sample is shown in Figure 4 (with corresponding representative HR-TEM images included in Figure S4). In summary, the mode peak size for the non-irradiated, ns-irradiated, and fs-irradiated samples are 2.51 nm, 2.44 nm, 2.22 nm, respectively. While the AuNPs are trending towards smaller diameters with decreasing pulse duration, the difference becomes more apparent upon the analysis of the population of AuNPs greater than 3.5 nm. For each sample, this region is represented by the green shaded area in Figure 4A–C. For the non-irradiated polymersome sample and the ns-irradiated polymersome sample, the percentage of AuNPs greater than 3.5 nm is approximately 13%, whereas this region only represents 7% of the total population for the fs-irradiated polymersome sample. This suggests that a larger population of nanoparticles undergo size reduction in response to fs irradiation which is consistent with findings by Delfour et al., in that larger particles will preferentially undergo fragmentation under fs irradiation [28].

Figure 4D,E shows the data overlay of the irradiated and non-irradiated samples for both ns and fs pulse durations, respectively. Both irradiated distributions are significantly different from the non-irritated particles at a level of 0.05 using a parametric z-test and log transformed data. The green triangle highlights the size region of interest where particle fragmentation is apparent. Concurrently, the opposite trend is seen for AuNPs within a smaller size range. The amount of AuNPs which fall below 2.5 nm trend towards larger counts for the fs-irradiated sample in comparison to the control sample, whereas this trend is not visible for ns irradiation. This suggests that pulse duration plays a significant role in vesicle rupture mechanism. In the case of ns irradiation, thermal relaxation through electron–phonon coupling begins to occur before the pulse delivery has been completed. Thus, thermal energy dissipation predominates and minimal fragmentation will occur. However, in the case of fs irradiation, the entirety of the pulse is delivered before energy can be dissipated through thermal pathways. Thus, this high energy density results in nanoparticle fragmentation.

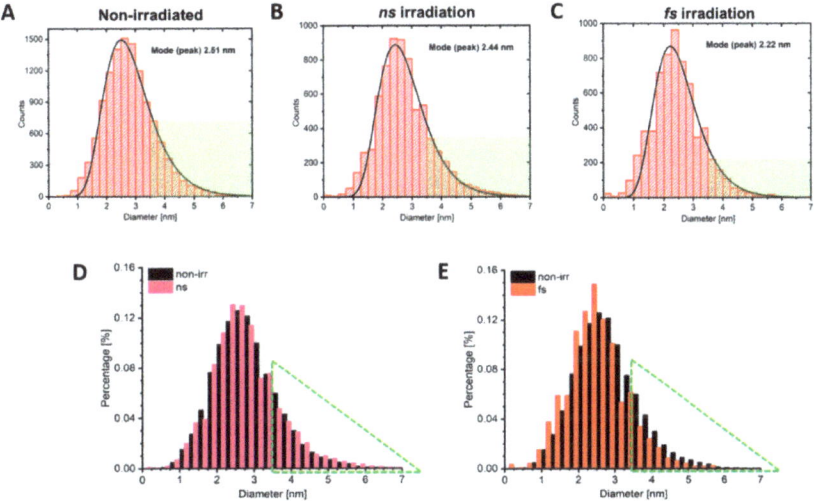

Figure 4. AuNP size distributions following vesicle irradiation. (**A–C**) The diameters of AuNP populations within polymersome membranes were measured without irradiation, and after ns and fs irradiation, respectively. The green shaded area highlights NPs with diameters greater than 3.5 nm. (**D,E**) Comparative population overlays for irradiated vs. non-irradiated samples. The green triangle highlights the size region of interest where particle fragmentation is apparent.

3.3. Controlling Complete Rupture vs. Pore Formation with Decreased Pulse Energy

The rupture frequency plots in Figure 3 show that irradiation below the rupture threshold does not result in complete rupture and, hence, the vesicles remain visibly intact via fluorescence microscopy. To probe membrane integrity and permeability at sub threshold levels, the polymersome cores were loaded with a hydrophilic fluorescent dye (FITC-dextran). It was observed that irradiation at these lower pulse energies leads to the release of the aqueous contents over a finite timescale as opposed to the instantaneous release event experienced with rupture. This behavior suggests that the membrane has undergone a poration event. At pulse energies with a low likelihood of rupture, the overall spherical shape of the vesicle can remain intact with little to no perceivable buckling and/or distortion of the membrane. Figure 5A shows a time series of fluorescent images of such an event.

In order to examine release characteristics of the porated vesicles, a series of fluorescent images were taken at regular intervals (10 s) that began approximately 3 s after the arrival of the incident laser pulse. The temporal release of encapsulant was determined by selecting a region of interest (ROI) both within the targeted polymersome and in the background (to account for possible photobleaching due to imaging post-irradiation) and then normalizing that value to the pre-irradiated intensity (see example calculation in Table S1). Results for fs irradiation are shown in Figure 5B where a minimum of three (non-rupturing) vesicles were measured at each pulse energy and graphed as the mean, with error bars denoting the standard error of the mean (SEM). Given that a rupture frequency of approximately 50% is observed for pulse energies at 210 nJ, this served as the upper energy boundary for the poration study. As shown in Figure 5B, the fluorescence intensity decreases in an exponential manner for all energies except at the lower boundary of 32 nJ where the membrane presumably reseals on a timescale equal to or shorter than the initial time step (i.e., 3 s). For the case of pulse energies near the upper boundary (140–220 nJ), encapsulant release occurs rapidly with a large degree (>80%) of the vesicle contents escaping (Figure 5C). Energies in the middle of the range (50–110 nJ) result in steady release with the removal of 50–60% of the contents. Given that the fluorescence continues to diminish up to the endpoint, it can be inferred that these pores are stable (i.e., did not grow boundless to rupture, or close) during the 120 s timeframe. Irradiation with pulse energies at the lower boundary (32 nJ) leads to a small reduction (13%) in the fluorescence signal which then quickly levels-off, indicating a short-lived pore that quickly self-heals. Note, attempts to observe temporal release under ns irradiation were erratic which indicates the formation of unstable pores, and hence were not included.

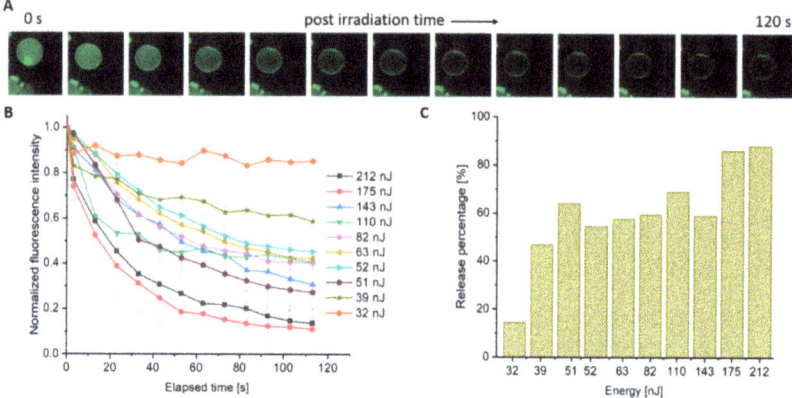

Figure 5. Encapsulant release as a function of time from Polymersomes loaded with 0.04% (*w/v*) AuNPs. (**A**) A sample time series displaying a decrease in fluorescence intensity as a function of time as FITC-dextran diffuses out of the vesicle after the delivery of a single 212 nJ pulse. (**B**) The fluorescence intensity inside of an individual vesicle is measured as a function of time following a single 532 nm, fs pulse at the indicated energy. Each line represents the fluorescence decay in a region of interest (ROI) (3–4 replicates +/− standard error of the mean (SEM)). (**C**) Total encapsulant release after 120 s at the indicated pulse energy.

To confirm that plasmonic excitation of the AuNP was playing an important role in pore formation, a control experiment was performed with polymersomes assembled without nanoparticles. While poration can still be observed in these empty membrane vesicles, a much higher pulse energy is required (Figure S5). Exposure of polymersomes without AuNPs to a 532 nm, 200 nJ, fs pulse resulted in <20% total encapsulant release. Compared to a membrane with AuNPs, the release percentage is near 90% with a rupture frequency close to 50%.

An interesting discrepancy was observed between energy threshold values for rupture and poration studies when hydrophilic encapsulant, FITC-dextran, was incorporated into the aqueous lumen. There is some precedent in the literature that the encapsulation of varying concentrations and molecular weights of dextrans into the aqueous lumen of photoresponsive polymersomes can alter vesicle deformation frequency [16]. We investigated the extent of the impact of this interaction on our rupture study by evaluating micron-sized PBD_{35}-*b*-PEG_{20} polymersomes formed both with and without FITC-dextran in the aqueous core. Rupture thresholds were determined in response to an fs pulse for samples containing and excluding AuNPs. When juxtaposed to Figure 3, Figure S6 shows that both vesicles prepared with and without AuNPs are observed to undergo complete rupture at a lower pulse energy when they do not contain FITC-dextran in their aqueous lumen. This result of decreased photosensitivity with dextran encapsulation is opposite to the trend found by Kamat et al., demonstrating an increase in photosensitivity with dextran incorporation. The authors observed an increase in deformation frequency and a decrease in elastic modulus, for two different PBD-*b*-PEG diblock copolymers, that scaled with the dextran molecular weight. The mechanism put forth states that the dextran penetrates the hydrophilic corona and reduces interfacial tension, thereby increasing the area of the inner leaflet. This difference in surface area can result in asymmetrical stress in the bilayer and destabilize the membrane. However, this reported effect was weaker for lower molecular weight dextrans, such as that conjugated to the FITC dye in this study (3–5000 Da). Several important distinctions between this work and that in the Kamat study are: (1) the inclusion of AuNPs in the membrane, (2) the molecular weight of the dextran used and (3) the use of ultrafast irradiation instead of CW. Further investigation is warranted to clarify the role played by dextrans in various photoresponsive polymersome systems.

3.4. Role of AuNP Concentration on Vesicle Poration

To investigate the effect of AuNP concentration on polymersome photo-response, vesicles with varying amounts of AuNPs loaded in the membrane were prepared. Polymersome samples with no AuNPs, 0.013%, 0.026%, 0.040%, 0.053%, and 0.066% (*w/v*) AuNPs were subjected to a single, 144 nJ fs pulse (chosen as a mid-range poration energy) and the subsequent cargo diffusion rate and total release were quantified for each sample. A distinct difference in the amount of fluorophore released from a minimally loaded sample with 0.013% AuNPs compared to a maximally loaded sample with 0.066% AuNPs could be observed, and is displayed in Figure 6A. In the case of the minimally loaded sample, there is negligible change in the fluorescence intensity in the vesicle core. However, in the case of the maximally loaded sample, a significant amount of the fluorophore diffused from the vesicle after two minutes. Using identical methods as those used to acquire data in Figure 5B, Figure 6A displays normalized fluorescence intensity as a function of time for a minimum of three vesicles at six distinct AuNP concentrations. Encapsulant release ranges from 90–100% (in the case of 0.066% (*w/v*) AuNPs) to 0% (in the case of 0% *w/v* AuNPs), scaling with AuNP concentrations.

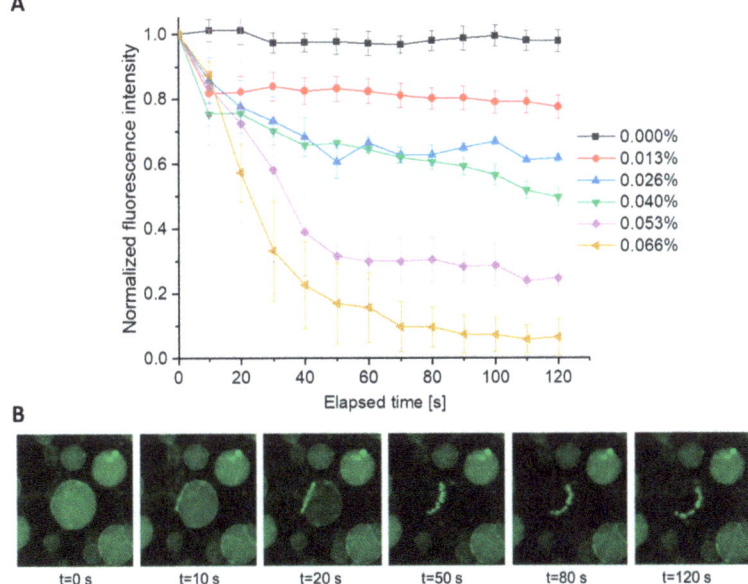

Figure 6. Effect of AuNP concentration on encapsulant release. (**A**) The fluorescence intensity inside of an individual vesicle is measured as a function of time following a single 532 nm, fs pulse (144 nJ). Each line represents the fluorescence decay in an ROI (3–4 replicates +/− SEM) with varying concentrations of AuNPs encapsulated within the polymersome membrane. (**B**) Poration leading to structural instability. A polymersome loaded with 0.066% (*w/v*) AuNPs is subjected to a single fs pulse (144 nJ). Upon irradiation, a large pore can be observed. The vesicle remains intact for 10–20 s following the pulse until the vesicle structure is compromised after approximately 50 s.

Interestingly, in a small population of maximally loaded vesicles (0.066% w/v AuNPs), structural changes resulting in vesicle destabilization were observed. This occurrence is distinctly different from the case of complete rupture which occurs nearly instantaneously upon pulse delivery. As shown in Figure 6B, the vesicle is destabilized as a function of time concurrent with cargo diffusion. This structural instability mostly occurs when the ratio of pore to vesicle size reaches a critical value at which the pore is unable to remain stable or self-heal, resulting in complete rupture.

4. Discussion

The ability to transport molecules in an unreacted state has significant importance for a variety of applications, namely, microreactors and drug delivery. Polymersomes have proven to be a promising candidate for such applications due to their robust nature and tunability. The self-assembly of PEO_{20}-b-PBD_{35} yields vesicles which are inherently are unreactive to stimuli, but in the work presented here, the incorporation of nanoparticles into the hydrophobic region of the membrane has proven to efficiently photosensitize the system in response to short pulse irradiation. The threshold energy for complete rupture was shown to reduce by a factor of five under fs irradiation which increased to a factor of 14 for ns irradiation.

By changing the pulse energy, cargo release can be precisely controlled ranging from instantaneous delivery, i.e., complete rupture, to regulated release over a timescale from seconds to minutes through pore formation. Additionally, the concentration of AuNPs loaded in the hydrophobic region of the membrane plays a role in cargo diffusion rates. This is significant for future biological applications where limiting the amount of AuNPs in the system may be favored.

The likely mechanism for membrane disruption is nanobubble formation around the excited nanoparticles upon thermal relaxation. In the case of ns irradiation, there is sufficient time for thermal transport which maximizes bubble formation. Such bubbles disrupt overall membrane stability resulting in complete rupture. This is indicated by the erratic nature of pore formation under ns irradiation. As evidenced by the size distributions of pre- and post-irradiated nanoparticles that have been incorporated into the membrane, those subjected to ns irradiation did not show signs of particle fragmentation while indications of size reduction are present in response to fs irradiation. The process of fragmentation decreases the localized thermal energy available for nanobubble formation, thus resulting in smaller regions of disruption and more stable pores.

This fundamental study has the potential to influence future work in the field of microreactors and drug delivery, where spatiotemporal control over cargo release is of central importance; however, implementation of the system in drug delivery applications will require scaling down to the more biologically appropriate nano-regime. Additionally, anisotropic gold nanoparticles may prove useful in shifting the response wavelength to the near-IR region of the spectrum for deeper tissue penetration in vivo. Work is currently underway to translate these findings from the micron- to the nano-regime.

Supplementary Materials: The following are available online at http://www.mdpi.com/2073-4360/12/10/2381/s1, Figure S1: Gel-assisted rehydration method for polymersome preparation, Figure S2: Ultrafast (fs) laser-microscope setup, Figure S3: Spot size determination, Figure S4: Representative HR-TEM images of nano-polymersomes containing AuNPs after irradiation, Figure S5: Release profile of encapsulant from polymersomes without AuNPs, Figure S6: Rupture threshold for vesicles without FITC-dextran in the core, Table S1: Example calculation for normalized intensity as shown in Figure 5, Figure 6 and Figure S5.

Author Contributions: Conceptualization, J.C.G. and S.M.O.; methodology, J.C.G. and S.M.O.; software, N/A; validation, G.M.D., A.R.R.; formal analysis, J.C.G., S.M.O., G.M.D., A.R.R., M.S.A.; investigation, G.M.D., A.R.R., E.R.H.; resources, J.C.G., S.M.O.; data curation, J.C.G., S.M.O.; writing—J.C.G.; writing—review and editing, J.C.G., S.M.O., A.R.R., G.M.D., M.S.A., E.R.H.; visualization, J.C.G., S.M.O., A.R.R., G.M.D.; supervision, J.C.G., S.M.O.; project administration, J.C.G., S.M.O.; funding acquisition, J.C.G., S.M.O. All authors have read and agreed to the published version of the manuscript.

Funding: This research was funded by NSF Award # 1531783, Rutgers University Provost's Catalyst Grant, and the NASA NJ Space Grant Consortium.

Acknowledgments: We thank Grace Herdelin for her early contributions to this work.

Conflicts of Interest: The authors declare no conflict of interest. The funders had no role in the design of the study; in the collection, analyses, or interpretation of data; in the writing of the manuscript, or in the decision to publish the results.

References

1. Sercombe, L.; Veerati, T.; Moheimani, F.; Wu, S.Y.; Sood, A.K.; Hua, S. Advances and challenges of liposome assisted drug delivery. *Front. Pharmacol.* **2015**, *6*, 286. [CrossRef] [PubMed]
2. Discher, B.M.; Won, Y.Y.; Ege, D.S.; Lee, J.C.; Bates, F.S.; Discher, D.E.; Hammer, D.A. Polymersomes: Tough vesicles made from diblock copolymers. *Science* **1999**, *284*, 1143–1146. [CrossRef] [PubMed]
3. LoPresti, C.; Lomas, H.; Massignani, M.; Smart, T.; Battaglia, G. Polymersomes: Nature inspired nanometer sized compartments. *J. Mater. Chem.* **2009**, *19*, 3576–3590. [CrossRef]
4. Rideau, E.; Dimova, R.; Schwille, P.; Wurm, F.R.; Landfester, K. Liposomes and polymersomes: A comparative review towards cell mimicking. *Chem. Soc. Rev.* **2018**, *47*, 8572–8610. [CrossRef] [PubMed]
5. Che, H.; Hest, J. Stimuli-responsive polymersomes and nanoreactors. *J. Mater. Chem. B* **2016**, *4*, 4632–4647. [CrossRef]
6. Smith, A.M.; Mancini, M.C.; Nie, S. Bioimaging: Second window for in vivo imaging. *Nat. Nanotechnol.* **2009**, *4*, 710–711. [CrossRef]
7. Cabane, E.; Malinova, V.; Meier, W. Synthesis of photocleavable amphiphilic block copolymers: Toward the design of photosensitive nanocarriers. *Macromol. Chem. Phys.* **2010**, *211*, 1847–1856. [CrossRef]
8. Dong, J.; Zeng, Y.; Xun, Z.; Han, Y.; Chen, J.; Li, Y.-Y.; Li, Y. Stabilized vesicles consisting of small amphiphiles for stepwise photorelease via uv light. *Langmuir* **2012**, *28*, 1733–1737. [CrossRef]
9. Schumers, J.-M.; Fustin, C.-A.; Gohy, J.-F. Light-responsive block copolymers. *Macromol. Rapid Commun.* **2010**, *31*, 1588–1607. [CrossRef]
10. Mabrouk, E.; Cuvelier, D.; Brochard-Wyart, F.; Nassoy, P.; Li, M.-H. Bursting of sensitive polymersomes induced by curling. *Proc. Natl. Acad. Sci. USA* **2009**, *106*, 7294–7298. [CrossRef]
11. Robbins, G.P.; Jimbo, M.; Swift, J.; Therien, M.J.; Hammer, D.A.; Dmochowski, I.J. Photoinitiated destruction of composite porphyrin-protein polymersomes. *J. Am. Chem. Soc.* **2009**, *131*, 3872–3874. [CrossRef] [PubMed]
12. Griepenburg, J.C.; Sood, N.; Vargo, K.B.; Williams, D.; Rawson, J.; Therien, M.J.; Hammer, D.A.; Dmochowski, I.J. Caging metal ions with visible light-responsive nanopolymersomes. *Langmuir* **2015**, *31*, 799–807. [CrossRef] [PubMed]
13. Kamat, N.P.; Liao, Z.; Moses, L.E.; Rawson, J.; Therien, M.J.; Dmochowski, I.J.; Hammer, D.A. Sensing membrane stress with near ir-emissive porphyrins. *Proc. Natl. Acad. Sci. USA* **2011**, *108*, 13984–13989. [CrossRef] [PubMed]
14. Np, K.; Gp, R.; Js, R.; Mj, T.; Ij, D.; Da, H. A generalized system for photo-responsive membrane rupture in polymersomes. *Adv. Funct. Mater.* **2010**, *20*, 2588–2596. [CrossRef]
15. Amstad, E.; Kim, S.-H.; Weitz, D.A. Photo- and thermoresponsive polymersomes for triggered release. *Angew. Chem* **2012**, *51*, 12499–12503. [CrossRef]
16. Zhang, H.; Cui, W.; Qu, X.; Wu, H.; Qu, L.; Zhang, X.; Mäkilä, E.; Salonen, J.; Zhu, Y.; Yang, Z.; et al. Photothermal-responsive nanosized hybrid polymersome as versatile therapeutics codelivery nanovehicle for effective tumor suppression. *Proc. Natl. Acad. Sci. USA* **2019**, *116*, 7744–7749. [CrossRef]
17. Miranda, D.; Lovell, J.F. Mechanisms of light-induced liposome permeabilization. *Bioeng. Transl. Med.* **2016**, *1*, 267–276. [CrossRef]
18. Mathiyazhakan, M.; Wiraja, C.; Xu, C. A concise review of gold nanoparticles-based photo-responsive liposomes for controlled drug delivery. *Nano-Micro Lett.* **2018**, *10*, 1–10. [CrossRef]
19. Mathiyazhakan, M.; Yang, Y.; Liu, Y.; Zhu, C.; Liu, Q.; Ohl, C.-D.; Tam, K.C.; Gao, Y.; Xu, C. Non-invasive controlled release from gold nanoparticle integrated photo-responsive liposomes through pulse laser induced microbubble cavitation. *Colloids Surf. B* **2015**, *126*, 569–574. [CrossRef]
20. Wu, G.; Mikhailovsky, A.; Khant, H.A.; Fu, C.; Chiu, W.; Zasadzinski, J.A. Remotely triggered liposome release by near-infrared light absorption via hollow gold nanoshells. *J. Am. Chem. Soc.* **2008**, *130*, 8175–8177. [CrossRef] [PubMed]
21. Lee, J.C.M.; Bermudez, H.; Discher, B.M.; Sheehan, M.A.; Won, Y.Y.; Bates, F.S.; Discher, D.E. Preparation, stability, and in vitro performance of vesicles made with diblock copolymers. *Biotechnol. Bioeng.* **2001**, *73*, 135–145. [CrossRef] [PubMed]

22. Greene, A.C.; Sasaki, D.Y.; Bachand, G.D. Forming giant-sized polymersomes using gel-assisted rehydration. *J. Vis. Exp.* **2016**, e54051. [CrossRef]
23. De Angelis, R.; Venditti, I.; Fratoddi, I.; De Matteis, F.; Prosposito, P.; Cacciotti, I.; D'Amico, L.; Nanni, F.; Yadav, A.; Casalboni, M.; et al. From nanospheres to microribbons: Self-assembled eosin y doped pmma nanoparticles as photonic crystals. *J. Colloid Interface Sci.* **2014**, *414*, 24–32. [CrossRef]
24. Habel, J.; Ogbonna, A.; Larsen, N.; Cherré, S.; Kynde, S.; Midtgaard, S.R.; Kinoshita, K.; Krabbe, S.; Jensen, G.V.; Hansen, J.S.; et al. Selecting analytical tools for characterization of polymersomes in aqueous solution. *RSC Adv.* **2015**, *5*, 79924–79946. [CrossRef]
25. Metwally, K.; Mensah, S.; Baffou, G. Fluence threshold for photothermal bubble generation using plasmonic nanoparticles. *J. Phys. Chem. C* **2015**, *119*, 28586–28596. [CrossRef]
26. Boulais, E.; Lachaine, R.; Hatef, A.; Meunier, M. Plasmonics for pulsed-laser cell nanosurgery: Fundamentals and applications. *J. Photochem. Photobiol. C* **2013**, *17*, 26–49. [CrossRef]
27. Klein-Wiele, J.H.; Simon, P.; Rubahn, H.G. Size-dependent plasmon lifetimes and electron-phonon coupling time constants for surface bound na clusters. *Phys. Rev. Lett.* **1998**, *80*, 45–48. [CrossRef]
28. Delfour, L.; Itina, T.E. Mechanisms of ultrashort laser-induced fragmentation of metal nanoparticles in liquids: Numerical insights. *J. Phys. Chem. C* **2015**, *119*, 13893–13900. [CrossRef]

Publisher's Note: MDPI stays neutral with regard to jurisdictional claims in published maps and institutional affiliations.

 © 2020 by the authors. Licensee MDPI, Basel, Switzerland. This article is an open access article distributed under the terms and conditions of the Creative Commons Attribution (CC BY) license (http://creativecommons.org/licenses/by/4.0/).

Article

Effect of Ionic Radius in Metal Nitrate on Pore Generation of Cellulose Acetate in Polymer Nanocomposite

Woong Gi Lee [1], Younghyun Cho [2,*] and Sang Wook Kang [1,3,*]

1. Department of Chemistry, Sangmyung University, Seoul 03016, Korea; even004@naver.com
2. Department of Energy Systems Engineering, Soonchunhyang University, Asan 31538, Korea
3. Department of Chemistry and Energy Engineering, Sangmyung University, Seoul 03016, Korea
* Correspondence: yhcho@sch.ac.kr (Y.C.); swkang@smu.ac.kr (S.W.K.); Tel./Fax: +82-2-2287-5362 (S.W.K.)

Received: 27 March 2020; Accepted: 21 April 2020; Published: 23 April 2020

Abstract: To prepare a porous cellulose acetate (CA) for application as a battery separator, $Cd(NO_3)_2·4H_2O$ was utilized with water-pressure as an external physical force. When the CA was complexed with $Cd(NO_3)_2·4H_2O$ and exposed to external water-pressure, the water-flux through the CA was observed, indicating the generation of pores in the polymer. Furthermore, as the hydraulic pressure increased, the water-flux increased proportionally, indicating the possibility of control for the porosity and pore size. Surprisingly, the value above 250 LMH (L/m^{2}h) observed at the ratio of 1:0.35 (mole ratio of CA: $Cd(NO_3)_2·4H_2O$) was of higher flux than those of CA/other metal nitrate salts ($Ni(NO_3)_2$ and $Mg(NO_3)_2$) complexes. The higher value indicated that the larger and abundant pores were generated in the cellulose acetate at the same water-pressure. Thus, it could be thought that the $Cd(NO_3)_2·4H_2O$ salt played a role as a stronger plasticizer than the other metal nitrate salts such as $Ni(NO_3)_2$ and $Mg(NO_3)_2$. These results were attributable to the fact that the atomic radius and ionic radius of the Cd were largest among the three elements, resulting in the relatively larger Cd of the $Cd(NO_3)_2$ that could easily be dissociated into cations and NO_3^- ions. As a result, the free NO_3^- ions could be readily hydrated with water molecules, causing the plasticization effect on the chains of cellulose acetate. The coordinative interactions between the CA and $Cd(NO_3)_2·4H_2O$ were investigated by IR spectroscopy. The change of ionic species in $Cd(NO_3)_2·4H_2O$ was analyzed by Raman spectroscopy.

Keywords: cellulose acetate; porosity; ionic radius; water-pressure

1. Introduction

Nanoporous materials have received much attention for their possibilities of being utilized for gas separation, catalysis, battery separators, medicine, and water treatment [1,2]. Currently, the preparation or synthesis of nanoporous materials has been explosively explored due to showing various and unique characteristics. For instance, the Lei Qian group suggested a special method to generate nanoparticles into porous materials with the evaporation phenomena of solvents [3]. On the other hand, the Dubinsky group showed that polymerization-induced phase separation could generate the hybrid porous material [4]. Recently, unique structures such as porous polymer networks (PPNs), porous organic polymers (POPs), metal-organic frameworks (MOFs) and porous aromatic frameworks (PAFs) have been investigated due to their gas storage characteristics [5–8]. On the other hand, Yiming et al. reported enhanced CO_2 uptake through microporous carbon material [9]. Additionally, Atsushi Takase et al. evaluated the SF_6 and N_2 separation ability of nanoporous carbons using an adsorption mechanism [10]. Moreover, porous materials as separators to improve the efficiency of batteries have been receiving attention. Specifically, porous organic materials have made promising candidates for highly efficient battery separators due to their easily tunable morphology by using smart functional materials and chemistry [11–14]. In detail, the lithium ion battery separator was prepared generally with polymers such as polyethylene (PE) or polypropylene (PP) because of their attractive physical and electrochemical stability. However, PP and PE separators have some drawbacks, which would show low wettability, resulting in the low ionic conductivity. Furthermore, low thermal stability played a role as an obstacle for practical applications. To overcome these drawbacks, porous inorganic material, nanoparticle, high thermal-resistance polymer and separator coated thermal material have been used as separator [15–17]. Especially, Meinan et al. developed the thermally stable separator with high performance by using the pure inorganic material [18]. Furthermore, the Lee group reported the thermally stable polyimide nanofiber with Al oxide particles as separator [19].

To date, porous materials have been formed through a variety of methods such as thermally induced phase separation, evaporation induced phase separation, phase inversion and track-etching [20–22]. For example, Jian Zhao et al. proposed a method to fabricate microporous material using the solvent evaporation-including separation. This method used a ternary solution, such as a liquid silicone rubber precursor of liquid paraffin and hexane, and this solution was cast to form a film. When the hexane and liquid paraffin was removed, micropores from less than 300 to over 1200 μm were generated in the silicone rubber film [23]. Hermsdorf et al. suggested the fabrication of porous polyimide film onto a silicon support coated with an oxide layer. Polyimide membranes are used as a starting material for a sample preparation based on the high energy of ion irradiation. As a result, the pore size in the membrane was an average of 73 nm and the pore shape was conical [24]. However, some problems of these methods are the complications and the high-cost of the process. To solve these drawbacks, our group reported the method to fabricate pores using the ionic liquid and solvated inorganic complex [25,26]. The nickel nitrate ($Ni(NO_3)_2 \cdot 6H_2O$) as inorganic complex was well dissolved in water or acetone and it can be applied to cellulose acetate solution dissolved acetone/water (w/w 8/2). As a result, the pore size and volume of the polymer matrix increased by water pressure and the pores were uniformly distributed [26]. In this study, we suggested the environment-friendly and energy-efficient method to synthesize the pores, which is close to the straight type in a cellulose acetate (CA) polymer matrix [27], using a combination of inorganic complex ($Cd(NO_3)_2 \cdot 4H_2O$) and isostatic water pressure. Especially, we investigated the effect of the ionic radius in metal nitrate on both the pore generation and pore control of cellulose acetate.

2. Experimental

Separator fabrication: polymer (cellulose acetate, Sigma-Aldrich Co. Milwaukee, WI, USA) was dissolved into acetone/water co-solvents. Then, cadmium (II) nitrate hexahydrate ($Cd(NO_3)_2 \cdot 4H_2O$, Sigma-Aldrich Co. Milwaukee, WI, USA) was incorporated into polymer solution by the mole ratio of the Cd salts to the monomeric unit of polymer. The prepared mixture was stirred for one hour at room temperature, and then casted for the free-standing film on the plate. Finally, the free-standing films were dried at atmosphere pressure for 30 min. The dried polymer films with $Cd(NO_3)_2 \cdot 4H_2O$ were placed into cells such that water molecules could be passed through the composite films. Then, the various water-pressures ranging from 2 to 8 bar were applied to the cells equipped with the polymer composites containing $Cd(NO_3)_2 \cdot 4H_2O$. The water flux of the films for varying porosity was measured and expressed as L/m^2h.

3. Results and Discussion

Scanning electron microscopy (SEM JSM-5600LV, JEOL, Tokyo, Japan) was utilized to observe the pore formation on and in polymers with the $Cd(NO_3)_2 \cdot 4H_2O$ additive after being exposed to external water-pressures. Figure 1a,b represents the surface morphology of the CA matrix which was dissolved in acetone/water co-solvents with the $Cd(NO_3)_2 \cdot 4H_2O$ additive. The SEM image clearly exhibited that there were abundant pores on the surface when the CA polymer with $Cd(NO_3)_2 \cdot 4H_2O$ was exposed to water-molecules, while the pores were not observed in the neat CA polymer even when the water-pressure was applied as in a previous study [28]. Furthermore, the cross-section images of the CA matrix containing the $Cd(NO_3)_2 \cdot 4H_2O$ additive shown in Figure 1c,d indicate that that pores were well fabricated in the polymers. In contrast, when the neat CA was dissolved in pure acetone, there were no pores on the surface of the polymer. In order to precisely control the pore size and porosity, we introduced the new method to utilize the $Cd(NO_3)_2 \cdot 4H_2O$ in acetone/water co-solvent. The resulting SEM image clearly implies that the pore size and porosity of polymer matrix could be controllable by inorganic salts and water-molecules. It was noted that there were two major plausible explanations for these phenomena: (1) the $Cd(NO_3)_2 \cdot 4H_2O$ aggregates solvated in the polymer matrix during the solidification while the volatile acetone rapidly evaporated, resulting in the remaining $Cd(NO_3)_2 \cdot 4H_2O$ aggregates forming well-defined pores in the CA polymer matrix; or (2) that there were ionic associations between the ions generated from the $Cd(NO_3)_2$ and the water molecules. These strong interactions delayed the evaporation of the water molecules in the CA polymer matrix, resulting in the formation of pores on the surface.

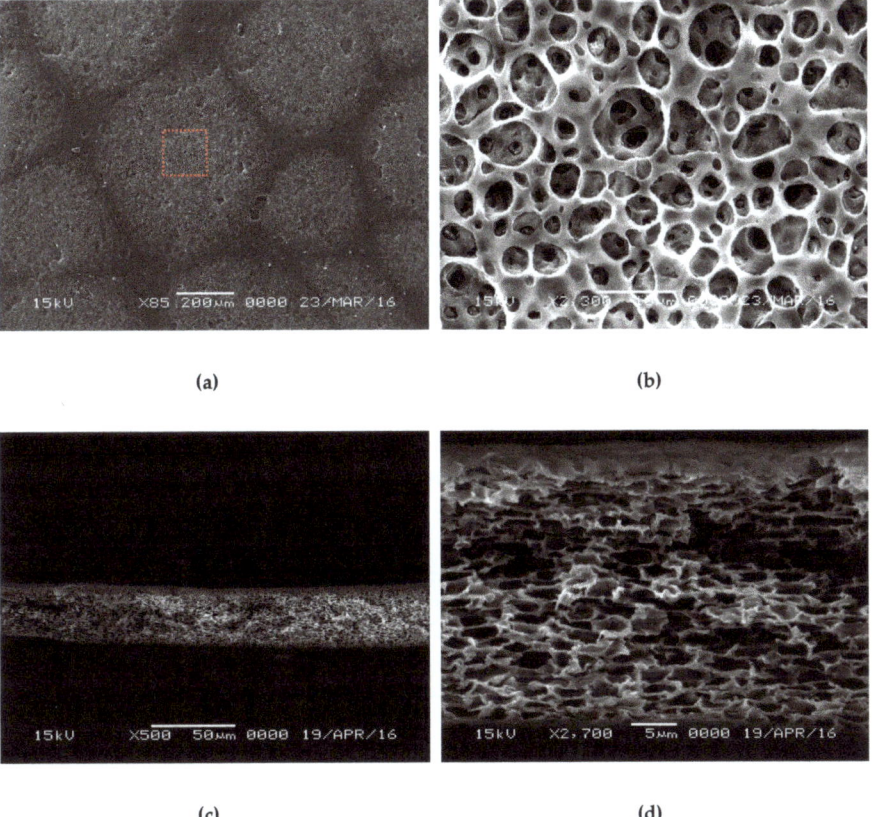

Figure 1. SEM images: (**a**) plane view of porous CA polymer exposed to external water-pressures (scale bar = 200 µm), (**b**) the enlarged red-dot square in plane view (scale bar = 10 µm), (**c**) cross-section view (scale bar = 50 µm) and (**d**) enlarged view of cross-section (scale bar = 5 µm).

The water flux through the porous CA polymer containing different ratios of $Cd(NO_3)_2 \cdot 4H_2O$ is shown in Figure 2. The water flux was not observed up to 2 bar, indicating that pores in CA polymer were not formed. However, the water flux gradually increased for 1:0.35 and 1:0.40 mol ratios CA/$Cd(NO_3)_2 \cdot 4H_2O$ polymer matrix at 3 bar. An explanation for these phenomena is that the water-channel was generated through weakened regions in the CA polymer, resulting in the observation of water-flux. However, the flux of neat CA polymer was not observed up to 7-bar water pressure and the low flux of 2.55 L/m^{2}h was observed only in the 8-bar water pressure [26]. From these results, it was thought that the CA polymer without Cd(NO$_3$)$_2$·4H$_2$O was confirmed not to have pores capable of being penetrated for water molecules through the polymer chains. Even though the amounts of Cd(NO$_3$)$_2$·4H$_2$O beyond the aforesaid ratios could be added, the increase in flux was not observed since the interactions between polymer chains were strengthened by crosslinking phenomena in CA polymer chains. Therefore, the best performance capable of being easily controllable for pores was the ratio of 1:0.35 for CA/Cd salts. Especially, the value above 250 LMH (L/m^2 h) observed at the ratio of 1:0.35 was higher flux than those of CA/other metal nitrate salts (Ni(NO$_3$)$_2$ and Mg(NO$_3$)$_2$) complexes. The higher value indicated that the larger and abundant pores were generated in the cellulose acetate with the same physical external forces. Thus, it could be thought that Cd(NO$_3$)$_2$·4H$_2$O salt played a role as a stronger plasticizing agent than the other metal nitrate salts such as Ni(NO$_3$)$_2$ and Mg(NO$_3$)$_2$.

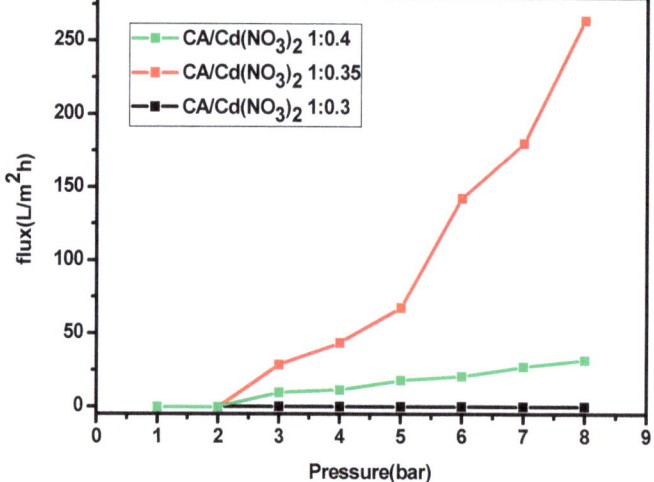

Figure 2. Flux measured through neat CA and CA with Cd(NO$_3$)$_2$·4H$_2$O at various water pressures.

Table 1 showed the comparison of atomic radius and ionic radius of Cd, Ni and Mg utilized in previous with this study to use Cd salts. As shown in Table 1, the atomic radius and ionic radius of Cd was largest among the three elements. Thus, it was thought that the Cd(NO$_3$)$_2$ that had relatively larger Cd could easily be dissociated into Cd ions and NO$_3^-$ ions. As a result, the free NO$_3^-$ ions could be readily hydrated with water molecules, resulting in the plasticization effect on the chains of cellulose acetate. Therefore, when the cellulose acetate containing the Cd(NO$_3$)$_2$ salts were exposed to high-intensive water-pressure, the pores could be generated more easily than in the other complexes containing Ni or Mg salts.

Table 1. Atomic radius and ionic radius of various elements utilized.

Elements	Atomic Radius (pm)	Ionic Radius (pm)
Cd	155	109
Ni	135	83
Mg	150	86
Zn	135	88

To understand the role of Cd(NO$_3$)$_2$·4H$_2$O solvates during the water pressure process, we performed Fourier transform infrared (FT-IR) spectra. Figure 3a shows the IR spectra of pristine polymer and 1/0.35 CA/Cd(NO$_3$)$_2$·4H$_2$O from 0 bar and 8 bar water pressures, respectively. The pristine polymer in IR spectra was observed at 3500 cm^{-1}, corresponding to the hydroxyl group of the CA polymer. In contrast, the CA polymer matrix with Cd(NO$_3$)$_2$·4H$_2$O showed the absorption bands for CA/Cd(NO$_3$)$_2$·4H$_2$O at 3400 cm^{-1}, resulting as the abundance of water molecules for Cd(NO$_3$)$_2$·4H$_2$O caused the intensity of the OH adsorption peak to be increased. The CA/Cd(NO$_3$)$_2$·4H$_2$O composites exposed to 8 bar were observed at the 3400 cm^{-1}, which decreased in intensity and shifted to 3500 cm^{-1}, indicating that a considerable amount of Cd(NO$_3$)$_2$·4H$_2$O was removed by high water-pressure process. The peak observed at the 1748 cm^{-1} as shown Figure 3b was shifted to 1731 cm^{-1} for polymer containing Cd(NO$_3$)$_2$·4H$_2$O. This result indicated that the carbonyl group of polymers interacted with the Cd ion. When the polymer matrix containing CA/Cd(NO$_3$)$_2$·4H$_2$O was exposed to external physical forces, most of the Cd(NO$_3$)$_2$·4H$_2$O was removed from the polymer matrix and recovered to 1748 cm^{-1}. Furthermore, the peak observed at the 1460 cm^{-1} as shown in Figure 3b disappeared when the polymer was exposed to external physical forces. It was also thought that the Cd(NO$_3$)$_2$·4H$_2$O was removed by physical pressure and the peak of 1460 cm^{-1} disappeared.

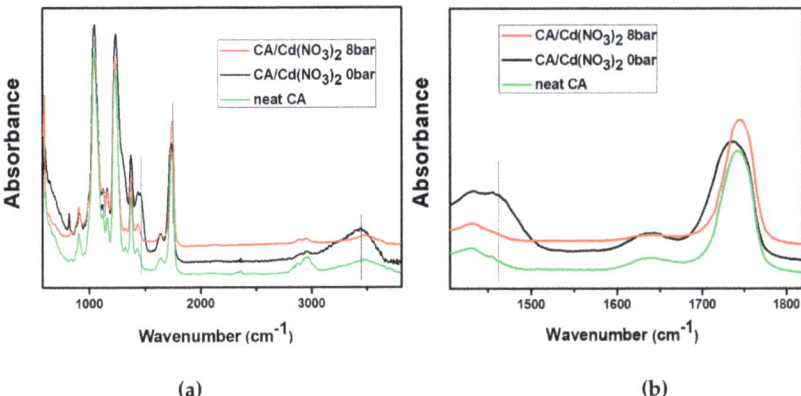

Figure 3. FT-IR spectra of neat CA and 1/0.35 CA/Cd(NO$_3$)$_2$·4H$_2$O polymer matrix at 0 bar and 8 bar water pressures: (**a**) total range and (**b**) enlarged specific region.

The thermal degradation pattern for polymers was observed by thermogravimetric analysis (TGA). Figure 4 showed that most of the pristine polymer and CA with Cd(NO$_3$)$_2$·4H$_2$O exposed at 8 bar were decomposed at around 300 °C. On the other hand, about 60 wt % of polymer with Cd(NO$_3$)$_2$·4H$_2$O at 0 bar was degraded at between 200 to 350 °C, and 40 wt % was degraded at between 350 to 800 °C. Since the boiling point of Cd(NO$_3$)$_2$·4H$_2$O was known to be 132 °C, it was thought that the chains in polymer were loosened by solvated cadmium nitrate. Thus, the loss of about 60 wt % CA with Cd(NO$_3$)$_2$·4H$_2$O was attributable to the degradation of solvated Cd(NO$_3$)$_2$·4H$_2$O and loosened polymer chain. However, the thermal stability of the polymer matrix increased after water-pressure treatment. The change in the thermal stability was generated as most of the Cd(NO$_3$)$_2$·4H$_2$O were removed in the polymer. As a result, the thermal stability of the polymer matrix (with Cd(NO$_3$)$_2$·4H$_2$O removed) by water-pressure treatment was similar to the neat polymer matrix.

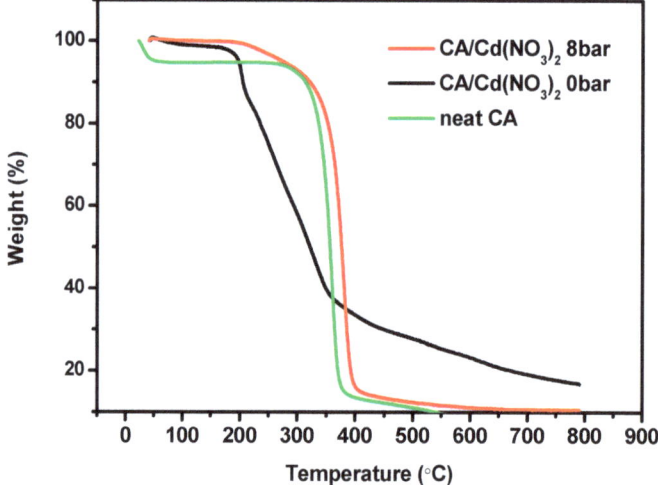

Figure 4. TGA analysis of neat CA and CA/Cd(NO$_3$)$_2$·4H$_2$O polymer matrix at 0 bar and 8 bar.

To investigate the ionic species such as ion aggregates, ion pairs and free ions in NO$_3^-$ ions for pristine Cd(NO$_3$)$_2$ and Cd(NO$_3$)$_2$ in the polymer, Raman spectra were acquired as shown in Figure 5. Since the ionic species of NO$_3^-$ at 1034 (free ions), 1040 (ion pairs) and 1045 (ionic aggregates) cm^{-1} was observed, the spectra were focused in the range of 1000 to 1080 cm^{-1}. As shown in Figure 5, the peak of pristine Cd(NO$_3$)$_2$ was observed at 1051.5 cm^{-1}.

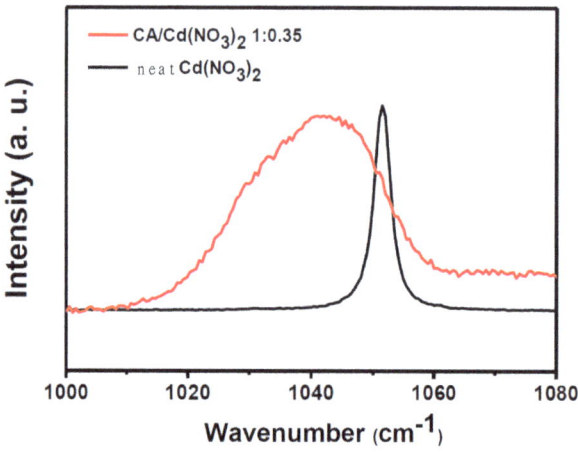

Figure 5. Raman spectra of neat Cd(NO$_3$)$_2$·4H$_2$O and CA/Cd(NO$_3$)$_2$·4H$_2$O polymer matrix.

When Cd(NO$_3$)$_2$ was incorporated to the polymer, the peak was shifted from 1051.5 to 1040 cm^{-1} and was broadened as shown Figure 5, indicating that the Cd(NO$_3$)$_2$ incorporated into the CA polymer existed as free ions and ion pairs as well as aggregates. In the case of the Cd(NO$_3$)$_2$ incorporated into the CA, the relative percentages for various ionic species were shown in Figure 6. Surprisingly, new peaks were observed at 1034.2 and 1045.6 cm^{-1}, indicating the free ions and ion aggregates. Deconvoluted results were shown for each of the regions in Figure 6.

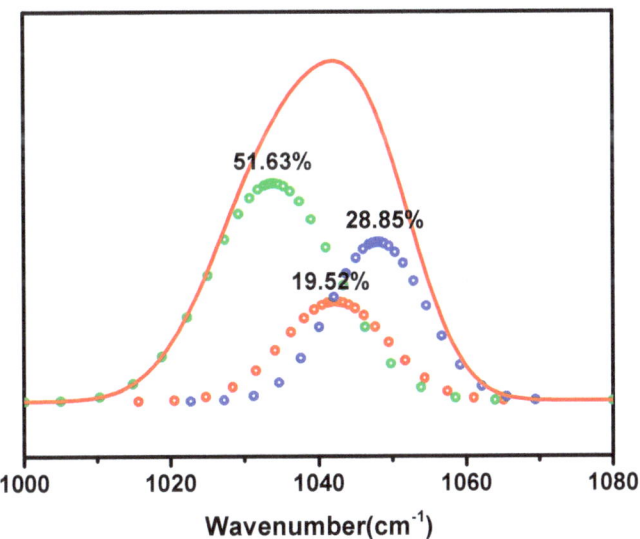

Figure 6. Deconvoluted Raman spectra for NO_3^- in the $CA/Cd(NO_3)_2 \cdot 4H_2O$ composite.

Based on these results, the $Cd(NO_3)_2$ incorporated into the polymer existed as abundant free ions in the polymer chain as shown in Table 2.

Table 2. Comparison of free ions, ion pair, and ionic aggregates % for NO_3^- in $Cd(NO_3)_2$ in CA.

	Free ions %	Ion Pair %	Ionic Aggregates %
$Cd(NO_3)_2$ in CA	51.63%	19.52%	28.85%

4. Conclusions

A porous polymer matrix was successfully fabricated by $Cd(NO_3)_2 \cdot 4H_2O$ and external physical forces as shown in Scheme 1. When water pressure as physical force was applied to the polymer matrix, pore size gradually increased. Furthermore, it was observed the water flux increased as the water pressure increased. These results indicated that the water molecule that penetrated into the polymer caused the chains to be weakened by solvated $Cd(NO_3)_2 \cdot 4H_2O$. Surprisingly, most of $Cd(NO_3)_2 \cdot 4H_2O$ in the polymer was extracted by the external forces. Furthermore, the relationship between radius characteristics in the metal salts and pore generation in the polymer was also investigated. It was found that since $Cd(NO_3)_2$ had relatively bigger Cd, it could easily be dissociated into Cd ions and NO_3^- ions and the free NO_3^- ions could be readily hydrated with water molecules, resulting in the plasticization effect on the chains of cellulose acetate. Therefore, when the cellulose acetate containing $Cd(NO_3)_2$ salts were exposed to high-intensive water-pressure, the pores could be more easily generated than other complexes containing Ni or Mg salts. Thus, these results were expected to be utilized for designing porous materials with specific pore-sizes.

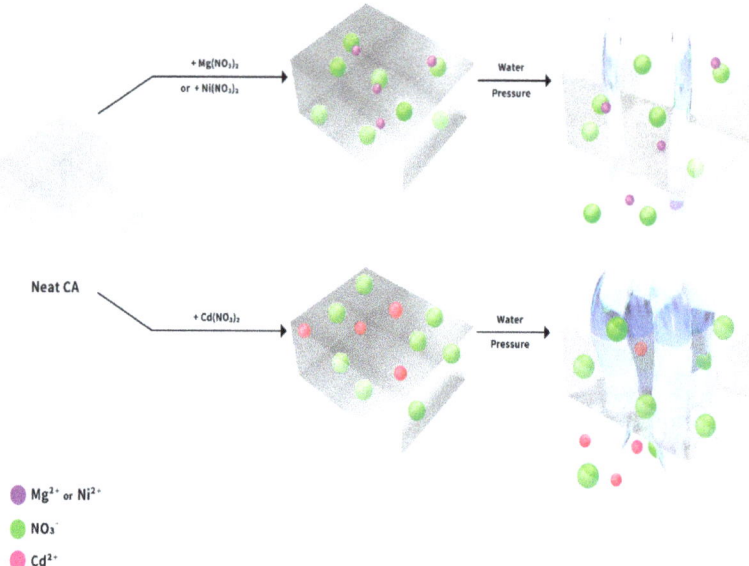

Scheme 1. Comparison of pore generation by species of metal salts in cellulose acetate.

Author Contributions: W.G.L. collected the experimental data and wrote the paper. Y.C. and S.W.K. analyzed the data and reviewed the paper. All authors have read and agreed to the published version of the manuscript.

Funding: This work was supported by the Basic Science Research Program (2017R1D1A1B03032583) through the National Research Foundation of Korea (NRF), funded by the Ministry of Science, ICT, and Future Planning. This work was also supported by the Soonchunhyang University Research Fund (No. 20181002).

Conflicts of Interest: The authors declare no conflict of interest.

References

1. Fehrmann, R.; Riisager, A.; Haumann, M. *Supported Ionic Liquids: Fundamentals and Applications*; Willey-VCH: Weinheim, Germany, 2014.
2. Prajapati, A.K.; Mondal, M.K. Hazardous As(III) removal using nanoporous activated carbon of waste garlic stem as adsorbent: Kinetic and mass transfer mechanisms. *Korean J. Chem. Eng.* **2019**, *36*, 1900–1914. [CrossRef]
3. Qian, L.; Ahmed, A.; Zhang, H. Formation of organic nanoparticles by solvent evaporation within porous polymeric materials. *Chem. Commun.* **2011**, *47*, 10001–10003. [CrossRef]
4. Dubinsky, S.; Petukhova, A.; Gourevich, I.; Kumacheva, E. Hybrid porous material produced by polymerization-induced phase separion. *Chem. Commun.* **2010**, *46*, 2578–2580. [CrossRef] [PubMed]
5. Kumar, P.; Vejerano, E.; Khan, A.; Lisak, G.; Ahn, J.H.; Kim, K.-H. Metal organic frameworks (MOFs): Current trends and challenges in control and management of air quality. *Korean J. Chem. Eng.* **2019**, *36*, 1839–1853. [CrossRef]
6. Li, W.; Zhang, G.; Zhang, C.; Meng, Q.; Fan, Z.; Gao, C. Synthesis of trinity metal-organic framework membranes for CO_2 capture. *Chem. Commun.* **2014**, *50*, 3214–3216. [CrossRef] [PubMed]
7. Zhu, J.-H.; Chen, Q.; Sui, Z.-Y.; Pan, L.; Yu, J.; Han, B.-H. Preparation and adsorption performance of crosslinked porous polycarbazoles. *Chem. Commun.* **2014**, *2*, 16181–16189.
8. Çalhan, A.; Deniz, S.; Romero, J.; Hasanoğlu, A. Development of metal organic framework filled PDMS/PI composite membranes for biobutanol recovery. *Korean J. Chem. Eng.* **2019**, *36*, 1489–1498. [CrossRef]

9. Zhang, Y.; Li, B.; Williams, K.; Gao, W.-Y.; Ma, S. A new microporous carbon material synthesized via thermolysis of a porous aromatic framework embedded with an extra carbon source for low-pressure CO_2 uptake. *Chem. Commun.* **2013**, *49*, 10269–10271. [CrossRef]
10. Takase, A.; Kanoh, H.; Ohba, T. Wide carbon nanopores as efficient sites for the separation of SF_6 from N_2. *Sci. Rep.* **2015**, *5*, 11994–11998. [CrossRef]
11. Xu, Q.; Kong, Q.; Liu, Z.; Zhang, J.; Wang, X.; Yue, R.L.L.; Cui, G. Polydopamine-coated cellulose microfibrillated membrane as high performance lithium-ion battery separator. *RSC Adv.* **2014**, *4*, 7845–7850. [CrossRef]
12. Lee, M.J.; Kim, J.H.; Lim, H.-S.; Lee, S.Y.; Yu, H.K.; Kim, J.H.; Lee, J.S.; Sun, Y.-K.; Guiver, M.D.; Suh, K.D.; et al. Highly lithium-ion conductive battery separators from thermally rearranged polybenzoxazole. *Chem. Commun.* **2015**, *51*, 2068–2071. [CrossRef]
13. Peng, K.; Wang, B.; Li, Y.; Ji, C. Magnetron sputtering deposition of TiO_2 particles on polypropylene separators for lithium-ion batteries. *RSC Adv.* **2015**, *5*, 81468–81473. [CrossRef]
14. Zhao, M.; Wang, J.; Chong, C.; Yu, X.; Wang, L.; Shi, Z. An electrospun lignin/polyacrylonitrile nonwoven composite separator with high porosity and thermal stability for lithium-ion batteries. *RSC Adv.* **2015**, *5*, 101115–101120. [CrossRef]
15. Sun, L.; Tang, S. Synthesis of bi-functionalized ionic liquid mesoporous alumina composite material and its CO_2 capture capacity. *Korean J. Chem. Eng.* **2019**, *36*, 1708–1715. [CrossRef]
16. Scrosati, B.; Hassoun, J.; Sun, Y.-K. Lithium-ion batteries. A look into the future. *Energy Environ. Sci.* **2011**, *4*, 3287–3295. [CrossRef]
17. Chung, S.-H.; Manthiram, A. High-performance Li-S batteries with an ultra-lightweight MWCNT-coated separator. *J. Phys. Chem. Lett.* **2014**, *5*, 1978–1983. [CrossRef]
18. He, M.; Zhang, X.; Jiang, K.; Wang, J.; Wang, Y. Pure inorganic separator for lithium ion batteries. *ACS Appl. Mater. Interfaces* **2015**, *7*, 738–742. [CrossRef]
19. Lee, J.; Lee, C.L.; Park, K.; Kim, I.D. Synthesis of an Al_2O_3-coated polyimide nanofiber mat and its electrochemical charateristics as a separator for lithium ion batteries. *Power Source* **2014**, *248*, 1211–1217. [CrossRef]
20. Lloyd, D.R.; Kim, S.S.; Kinzer, K.E. Microporous membrane formation via thermally-induced phase separation. II. Liquid-Liquid phase separation. *J. Membr. Sci.* **1990**, *52*, 239–250. [CrossRef]
21. Mo, D.; Liu, J.D.; Duan, J.L.; Yao, H.J.; Latif, H.; Cao, D.L.; Chen, Y.H.; Zhang, S.X.; Zhai, P.F.; Liu, J. Fabrication of different pore shapes by multi-step etching technique in ion-irradiated PET membranes. *J. Membr. Sci.* **2014**, *333*, 58–63. [CrossRef]
22. Apeal, P.Y.; Ramirez, P.; Blonskaya, I.V.; Orelovitch, O.L.; Sartowska, B.A. Accurate characterization of single track-etched, conical nanopores. *Chem. Commun.* **2014**, *16*, 15214–15223. [CrossRef] [PubMed]
23. Zhao, J.; Luo, G.; Wu, J.; Xia, H. Preparation of microporous silicone rubber membrane with tunable pore size via solvent evaporation-induced phase separation. *ACS Appl. Mater. Interfaces* **2013**, *5*, 2040–2046. [CrossRef] [PubMed]
24. Hermsdorf, N.; Sahre, K.; Volodin, P.; Stamm, M.; Eichhorn, K.J.; Cunis, S.; Gehrke, R.; Panagiotou, P.; Titz, T.; Muller-Buschbaum, P. Supported particle track eyched polyimide membranes: A grazing incidence small-angle X-ray scattering study. *Langmuir* **2004**, *20*, 10303–10310. [CrossRef] [PubMed]
25. Lee, W.G.; Hwang, J.; Kang, S.W. Control of nanoporous polymer matrix by an ionic liquid and water pressure for applications to water-treatment and separator. *J. Chem. Eng.* **2016**, *284*, 37–40. [CrossRef]
26. Lee, W.G.; Kim, D.H.; Jeon, W.C.; Kwak, S.K.; Kang, S.J.; Kang, S.W. Facile control of nanoporosity in Cellulose Acetate using Nickel(II) nitrate additive and water pressure treatment for highly efficient battery gel separators. *Sci. Rep.* **2017**, *7*, 1287–1295. [CrossRef]
27. Klemm, D.; Heublein, B.; Fink, H.P.; Bohn, A. Cellulose: Fascinating biopolymer and sustainable raw material. *Angew. Chem. Int. Ed. Engl.* **2005**, *44*, 3358–3393. [CrossRef]
28. Lee, W.G.; Kang, S.W. Control of Pore in Cellulose Acetate containing Mg salt by Water Pressure Treatment for Applications to Separators. *J. Ind. Eng. Chem.* **2019**, *70*, 103–106. [CrossRef]

© 2020 by the authors. Licensee MDPI, Basel, Switzerland. This article is an open access article distributed under the terms and conditions of the Creative Commons Attribution (CC BY) license (http://creativecommons.org/licenses/by/4.0/).

Article

3D Hierarchical, Nanostructured Chitosan/PLA/HA Scaffolds Doped with TiO$_2$/Au/Pt NPs with Tunable Properties for Guided Bone Tissue Engineering

Julia Radwan-Pragłowska [1], Łukasz Janus [1], Marek Piątkowski [1,*], Dariusz Bogdał [1] and Dalibor Matysek [2]

1. Department of Physical Chemistry, Faculty of Chemical Engineering and Technology, Cracow University of Technology, 31–155 Cracow, Poland; j.radwan@doktorant.pk.edu.pl (J.R.-P.); lukasz.janus@doktorant.pk.edu.pl (Ł.J.); pcbogdal@cyf-kr.edu.pl (D.B.)
2. Faculty of Mining and Geology, Technical University of Ostrava; 708 00 Ostrava, Czech Republic; dalibor.matysek@vsb.cz
* Correspondence: marek.piatkowski@pk.edu.pl; Tel.: +48-12-628-27-33

Received: 28 December 2019; Accepted: 25 March 2020; Published: 2 April 2020

Abstract: Bone tissue is the second tissue to be replaced. Annually, over four million surgical treatments are performed. Tissue engineering constitutes an alternative to autologous grafts. Its application requires three-dimensional scaffolds, which mimic human body environment. Bone tissue has a highly organized structure and contains mostly inorganic components. The scaffolds of the latest generation should not only be biocompatible but also promote osteoconduction. Poly (lactic acid) nanofibers are commonly used for this purpose; however, they lack bioactivity and do not provide good cell adhesion. Chitosan is a commonly used biopolymer which positively affects osteoblasts' behavior. The aim of this article was to prepare novel hybrid 3D scaffolds containing nanohydroxyapatite capable of cell-response stimulation. The matrixes were successfully obtained by PLA electrospinning and microwave-assisted chitosan crosslinking, followed by doping with three types of metallic nanoparticles (Au, Pt, and TiO$_2$). The products and semi-components were characterized over their physicochemical properties, such as chemical structure, crystallinity, and swelling degree. Nanoparticles' and ready biomaterials' morphologies were investigated by SEM and TEM methods. Finally, the scaffolds were studied over bioactivity on MG-63 and effect on current-stimulated biomineralization. Obtained results confirmed preparation of tunable biomimicking matrixes which may be used as a promising tool for bone-tissue engineering.

Keywords: smart hybrid materials; properties of nanoparticles–reinforced polymers; biotechnology

1. Introduction

Bone is one of the tissues with the ability of self-regeneration and constant remodeling [1,2]. Despite this fact, each year, over four million surgeries are performed in order to treat this tissue's defects, using autografts or bone substitutes [1,2]. The history of bone pathologies which cannot be self-healed is very long and concerns both congenital and acquired ones. These include traumas, neoplasm, infection or failed arthroplasty, spine arthrodesis, implant fixation, and others [2]. Because of that, bone is known to be the second most commonly replaced tissue [1,2]. Since the application of a patient's own bone is limited by the size of the defect, bone substitutes are rapidly gaining attention. Grafts coming from other donors, so called allografts, carry the risk of transmitting disease, infection, and its rejection. Alternatively, used metal (stainless steel, pure titanium, or its alloys) and synthetic biomaterials have some major limitations, such as poor biocompatibility, biocorrosion, and consumption over time. Non-invasive methods can only accelerate the healing process [2]. To overcome this issues, novel types of biomaterials are developed which mimic natural bone composition and structure [1–4].

Tissue engineering is one of the most powerful tools used in medicine which brings hope to the millions of patients suffering from various dysfunctions caused by diseases, traumas, or injuries. Its main objective is to restore, improve, or maintain biological functions of damaged tissue [5]. This strategy involves application of the mechanical support called scaffold, which also provides biological help to the cells, due to its chemical composition. To improve their properties, scaffolds can be prepared from various raw materials and are often functionalized with bioactive molecules, such as growth factors or antibiotics [1–3,5].

The biomaterials for bone regeneration are prepared in the form of three-dimensional spatial structure which mimics natural tissue by various methods. Porous materials can be embedded with various substances which promote osteoconduction. The ideal scaffold for this application should meet five conditions, namely biocompatibility, biodegradability, bioactivity, appropriate architecture, and high durability [4–8].

Scaffold dedicated to bone-tissue engineering should provide a temporary mechanical support for the damaged area and stimulate tissue regeneration [2,9]. Therefore, it must be characterized by highly porous architecture, to enable bone ingrowth, as well as neovascularization. The scaffold should constitute a template which promotes extracellular matrix formation and a place for osteoblasts to adhere and proliferate. Such biomaterials can have a form of nanofibers, hydrogels, metal alloys, β-TCP, HA powders, and granules or bioactive glasses [10,11]. Among them, the most promising ones are nanofibrous scaffolds, which can be prepared through electrospinning of the polymer solution since they biomimic natural ECM architecture [12–16]. Their properties, such as high surface area and porosity make them a highly desired materials in terms of bone-tissue-defect treatment [17–19]. However, due to the hierarchical nature of the bone, nanofibrous/porous composites seem to be the most promising ones [5,9].

Scaffolds for bone-tissue engineering that are prepared from minerals have poor mechanical durability, but at the same time, they exhibit good biocompatibility and the potential for various molecules' release. Natural polymers undergoing enzymatic biodegradation which speed rate may be adjusted are used as scaffold components and drug-delivery and -release systems [20]. The most common ones are collagen, fibrin, gelatin, chitin, and its derivatives, or alginate; additionally, biodegradable polymers, such as PLGA/PLA and PCL are widely used [21–24]. Notably, their low durability and fast degradability means that they must be used in conjunction with other materials [14–16]. Electrospinning enables preparation of micro- and nanofibers, using both raw polymers solution as well as composites and nanocomposites [19]. To enhance NFs properties, ready products may undergo postprocessing functionalization by various substances, such as hydroxyapatite, β-TCP, metal, or metal oxide nanoparticles [5,9,10,25–27]. Such an approach enables preparation of bone-tissue scaffolds with tunable properties, including programmed biodegradability rate, enhanced mechanical durability, crystallinity, antibacterial, bioactivity, and others [25,28]. There are some synthetic polymers approved by the FDA, such as polycaprolactone (PCL) and polylactic acid PLA, which can be successfully used for nanofibers' preparation. Other polymers include polyglycolic Acid (PGA) and polyethylene glycol (PEG). Nevertheless, their degradation products cause a local pH decrease due to their acidic nature, which negatively affects proteins and trigger inflammation [1–5]. Chitosan is a poly(saccharide) which is used for porous sponges and nanofibers preparation, yet due to its problems associated with solubility in different solvents, this biopolymer application requires chemical or physical modification [20–24].

The aim of the following research was to develop a novel hybrid material of hierarchical structure with enhanced biological properties. Prepared three-dimensional biomaterials for bone-tissue engineering were developed based on the application of Au [29,30], Pt [31], and TiO_2 [32–35] nanoparticles embedded on PLA nanofibers functionalized with hydroxyapatite and highly porous chitosan aerogel matrix. The addition of the nanoparticles has a positive impact on the mechanical properties of the scaffolds, and due to their ability of current conductivity, they enable cell-proliferation stimulation by direct current [25–28]. Moreover, the presence of the NPs has a positive impact on various cell responses by itself [25–27]. The nanoparticles can also help to reduce the risk of bacterial infections, since they exhibit antibacterial activity against various bacterial strains, such as S. aureus or E. coli. The bactericidal effect occurs due to genetic material damage or disruption of cell membrane [25–28]. The composites were prepared by using polymers of known applicability in tissue engineering [36–38]. Ready scaffolds with tunable properties were characterized over their physicochemical properties, such as chemical structure, crystallinity, morphology, effect on biomineralization, and biocompatibility with human osteosarcoma (MG-63) cell lines [39–41]. Obtained results showed that proposed biomaterials display extraordinary properties and may significantly contribute to bone-tissue-engineering development.

2. Materials and Methods

2.1. Materials

Fungal chitosan (300,000 g/mol) and 85% deacetylation degree were purchased from PolAura (Dywity, Poland). Poly(lactic acid) (1,500,000 g/mol, melting point = 210 ± 10 °C), acetone (analytical grade), NaCl, NH_3, H_3PO_4, $Ca(NO_3)_2 \cdot 6H_2O$, $NaHCO_3$, KCl, $K_2HPO_4 \cdot 3H_2O$, $MgCl_2 \cdot 6H_2O$, KOH, HCl, $CaCl_2$, chloroauric acid, Na_2SO_4, $(CH_2OH)_3CNH_2$, methanol, ethanol, XTT assay, human osteosarcoma MG-63 cell line, titanium (IV) butoxide, nitric acid, chloroplatinic acid, Dulbecco's Modified Eagle Medium (DMEM) with high glucose content, phosphate buffer solution (PBS), streptomycin/penicillin, and trypsin were purchased from Sigma-Aldrich, Poznań, Poland. All other reagents were of analytical grade, purchased from Sigma-Aldrich, Poznań, Poland.

2.2. Methods

2.2.1. Nanoparticles' Preparation and Characterization

Briefly, gold nanoparticles were prepared via reduction reaction of chloroauric acid, using sodium citrate at 90 °C and mixed for 1 h. Platinum nanoparticles were obtained by using chloroplatinic acid as a platinum source, which was instilled inside heated up ethylene glycol (150 °C) with constant mixing for two hours, until black color was reached. For the titanium (IV) oxide NPs obtainment, as a precursor, titanium (IV) butoxide was used. To obtain the nanoparticles, the precursor was mixed with ethanol for 30 min, followed by water ethanol solution addition. The ready white precipitate was washed out with distilled water and left to dry. The ready nanoparticles were purified by membrane dialysis, using MWCO 10,000–12,000 Da (Bionovo, Legnica, Poland). For the preparation of hydroxyapatite nanoparticles $Ca(NO_3)_2 \cdot 6H_2O$, ammonia and H_3PO_4 were used. The ready precipitate was washed out with methanol and calcinated at 250 °C for one hour. The resulted nanoparticles' polydispersity index (PDI) was calculated, using the Equation (1), given below, based on the TEM images:

$$PDI = SD/mean \qquad (1)$$

where PDI is polydispersity index, SD is standard deviation, and mean is the arithmetic mean.

2.2.2. Nanofibers Preparation

To obtain a homogenous solution of poly (lactic acid), previously purified polymer was dissolved in the acetone of analytical-grade purity. The solution concentration was 10%. To prepare nanofibers, Industrial Electrospinning System RT Advanced was applied (Linari NanoTech, Pisa, Italy). Nanofibers were prepared at two different potentials, namely 30 and 35 kV. The distance from the spindle was 10. The polymer solution was dosed with the constant speed of 10 mL per hour, with the rotation speed of 1000 RPM. The collector was coated, using aluminum foil. The needle diameter was 500 μm. Obtained nanofibers were left under room temperature, until complete solvent evaporation, and peeled of the foil. Dried nanofibrous mats were used for further investigations.

2.2.3. Chitosan Aerogels Preparation

To prepare an aerogel, for each sample, 0.5 of fungal chitosan was dissolved in aspartic acid water solution, until homogenous solution obtainment. Then, 5 mL of 1,2-propanodiol was added, and the vessel containing polymer was placed inside of the Prolabo Synthewave microwave reactor (Wrocław, Poland). The crosslinking reaction was carried out for 10 min. During the first step, water evaporation was noticed, followed by a crosslinking process. The ready product was washed out from unreacted amino acid residues, until pH = 7. Then, it was mixed with the HA NPs (10 wt %). The scaffolds were freeze-dried.

2.2.4. Three-Dimensional Scaffolds' Preparation

The hybrid scaffolds were prepared by placing nanofibrous 3D mat onto chitosan aerogel which contained previously absorbed water (5 mL), followed by lyophilization. The samples were frozen at −20 °C and lyophilized at 0.30 mPa, using ALFA lyophilizator (Donserv, Warszawa, Poland). Then, to embed metallic nanoparticles, 1% methanol solutions containing TiO_2, Au, and Pt NPs were instilled onto chitosan covered with PLA NFs. The modified biomaterials were left to dry. The chemical composition of the ach sample is given in the Table 1. Each sample contained 0.5 g of the chitosan and 10% w/w of hydroxyapatite. The concentration of the PLA solution for the fibers' preparation was 10%. As a solvent, acetone was used.

Table 1. Three-Dimensional biomaterials' composition.

Sample	NPs Type, %	Potential, kV
CS-PLA-30-HA	-	30
CS-PLA-35-HA	-	35
CS-PLA-30-HA-TiO_2	TiO_2, 1.0	30
CS-PLA-35-HA-TiO_2	TiO_2, 1.0	35
CS-PLA-30-HA-Au	Au, 1.0	30
CS-PLA-35-HA-Au	Au, 1.0	35
CS-PLA-30-HA-Pt	Pt, 1.0	30
CS-PLA-35-HA-Pt	Pt, 1.0	35

2.2.5. FT-IR Analysis

Fourier-Transform Infrared Spectroscopy (FT-IR) analysis was performed, using FT-IR Nexus 470 Thermo Nicolet spectrometer obtained from Thermo Fisher Scientific (Waltham, MA, USA). For the measurements, an ATR diamond adapter was used. For FT-IR spectra collecting, each sample was completely dried.

2.2.6. Liquid-Uptake Ability and Swelling Degree

Liquid-uptake studies were performed in simulated body fluid (SBF). To determine swelling capability, the weighed samples were immersed in SBF for 24 h and weighed again. Based on the weight change, swelling capacity (SC) was determined. To verify the swelling degree, the change in the dimensions of the scaffolds was studied. For this purpose, the samples were cut into cubes (1 cm × 1 cm × 1 cm) and swollen with SBF. After 5 min and 24, their dimensions were measured. Statistical analysis was performed by Excel software, and a $p < 0.05$ value was found to be statistically significant.

2.2.7. XRD Analysis

X-ray powder diffraction (XRD) was carried out, using BRUKER Advanced D8 (Zastávka, Czech Republic). For each analysis, a fully dried sample was used in the amount of 50 mg.

2.2.8. SEM and TEM Analysis

Nanoparticles morphology and diameter were evaluated by Transmission Electron Microscope (TEM), Jeol (Peabody, MA, USA). For the analysis, nanoparticles were dissolved in the methanol of analytical-grade purity, followed by placing onto formvar-coated Cu mesh and left to evaporate under room temperature. The analysis was carried out under HT = 80,000 V, exposure time of 800 ms, and electron (e) dose of 2771.9 e/nm^2. The nanoparticles' size was determined by Jeol software. The scaffolds' components were investigated via an FEI Quanta 650 FEG Scanning Electron Microscope purchased from FEI (ThermoFisher Scientific, Oregon, USA) with HV = 10 kV. The elemental analysis was performed by using an EDAX® adapter (X-ray fluorescence method).

2.2.9. Inorganic Matrix Formation and DC-Induced Inorganic Matrix Formation Study

To determine prepared scaffolds in terms of their positive effect on the inorganic matrix formation, which is an important part of the biomineralization, the scaffolds were placed inside sealed vials containing SBF and stored inside of a CO_2 incubator, at 37 °C, for 7 days. To evaluate the scaffolds' potential to electrostimulation of biomineralization process, samples were placed inside cubic glass vessels containing two platinum electrodes 5V (direct current) immersed in SBF. Then, the biomaterials were investigated over coverage with HA crystals, using Atomic Absorption Spectroscopy (AAS) PU-9100x (Philips), as well as SEM and XRF methods.

2.2.10. Cytotoxicity Study

Cytotoxicity study on a human osteosarcoma cell line MG-63 was carried out, using 2,3-Bis-(2-methoxy-4-nitro5-sulfophenyl)-2H-tetrazolium-5-carboxanilide salt assay (XTT). The salt underwent reduction reaction due to the presence of enzymes coming from metabolically active cells. The cytotoxicity study was carried out according to norm ISO 10993-5 Biological evaluation of medical devices, tests for in vitro cytotoxicity. For the experiments, an MG-63 cell line was used. Bone cells culture was conducted for 7 days, under typical conditions (95% CO_2, high humidity and 37 °C). The cell culture medium (DMEM) with high glucose content, 2 mM Glutamine supplemented with 10% fetal bovine serum was changed every two days. To perform XTT assay, UV-Vis spectrophotometer Agilent 8453 was applied (Santa Clara, CA, USA).

Statistical analysis was performed by Excel software, and a $p < 0.05$ value was found to be statistically significant.

3. Results and Discussion

3.1. FT-IR Analysis

The 3D hybrid scaffolds were prepared via combination of PLA nanofibers and chitosan crosslinked aerogel containing hydroxyapatite nanoparticles. The biomaterials were further modified with three types of nanoparticles: TiO_2, Au, and Pt. The general principle of the supporting matrixes is given in Figure 1.

The hybrid scaffolds' components were characterized by FT-IR method (crosslinked chitosan bottom layer and nanofibrous PLA top layer). Figure 2 shows that, during the chitosan crosslinking process, using L-aspartic acid in the 1,2-propanodiol environment, new bonds were formed between free amino groups and carboxyl groups of the amino acid, since the intensity of amide bonds (1658 cm^{-1} for pure chitosan) has significantly increased (1658 cm^{-1} for crosslinked chitosan).

The incorporation of the aspartic acid also proves the increased intensity of free amino groups in the crosslinked polymer (1574 and 1148 cm^{-1}), compared to pure chitosan (1593 and 1149 cm^{-1}). Such a modification pathway results in the maintenance of free amino groups which are responsible for many positive features and induce cellular responses [20–22]. FT-IR spectrum of the nanofibers prepared from the poly (lactic acid) exhibit typical for this polymer band of high intensity at 1751 cm^{-1}, and no bands coming from impurities or not-evaporated solvent can be spotted. The other bands come from C–H bending (1382 cm^{-1}) and C–O stretching (1182–1044 cm^{-1}) [37,38]. What is important is that raw PLA is known for its low bioactivity, as well as hydrophobicity. Thus, application of crosslinked chitosan containing hydrophilic groups can increase cells' adhesion and cellular responses [15]. Moreover, crosslinked polymer spectrum exhibits a band coming from carboxyl groups at 3217 cm^{-1} that may be present due to the surface degradation as a result of microwave-irradiation, as well as the aspartic acid molecules, which are not fully incorporated into the polymeric matrix only partially grafted leaving one-COOH group free. Superficial degradation of the chitosan causes very slight changes in its chemical structure, as a result of the oxidation of the hydroxyl groups, namely CH$_2$OH, which turns into COOH. Due to the presence of the oxygen in the reaction atmosphere, carboxyl groups are being formed, thus replacing hydroxyl ones. No other degradation products are observed. Such chemical structure will provide appropriate chemical conditions for bone-forming hydroxyapatite crystallization and stimulate its nucleation initiation, since polymeric layers will act as an organic template [40,41].

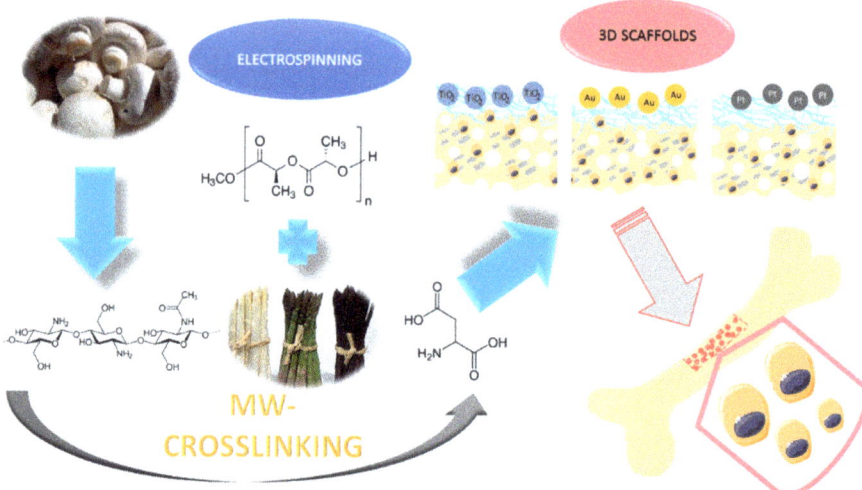

Figure 1. The general 3D scaffolds' obtainment strategy and application.

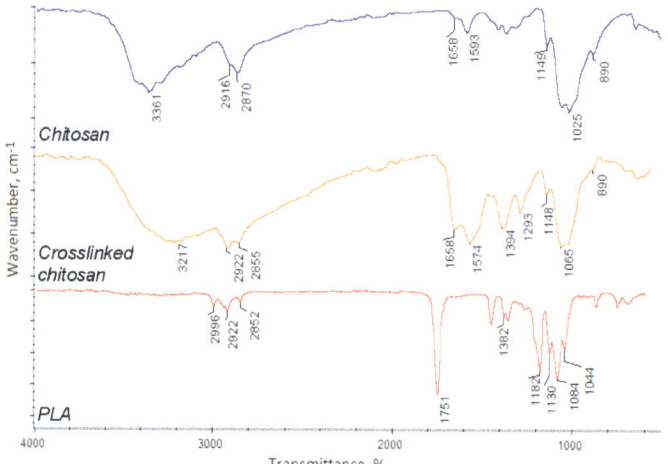

Figure 2. FT-IR spectra of the pure chitosan, crosslinked chitosan, and poly (lactic acid) nanofibers.

Based on the FT-IR data, the proposed chemical structure is given in Figure 3. Such hierarchical composition should well mimic natural bone. Acidic groups provide local charge accumulation, which helps calcium and phosphorous ions' binding [41]. The chitosan layer mimics the naturally occurring organic bone-forming components on which minerals deposition occurs [40]. Introduced molecules of aspartic acid, as well as poly (lactic acid), enable biological system imitation. Importantly, the chemical structure is rich in both anions and cations, which should help to maintain salts present in the simulated body fluids at elevated level at all times [40].

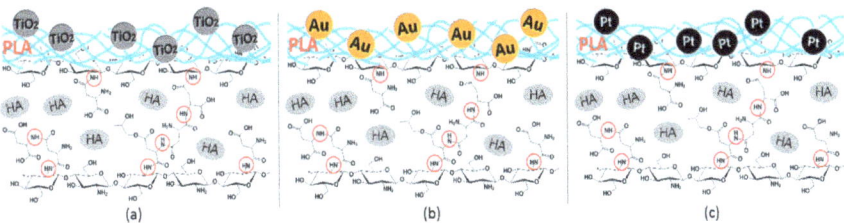

Figure 3. (**a**) 3D hybrid scaffolds' chemical structure doped with TiO_2 NPs; (**b**) 3D hybrid scaffolds chemical structure doped with Au NPs; and (**c**) 3D hybrid scaffolds doped with Pt NPs.

3.2. Swelling Properties of the 3D Scaffolds

Hydrogels, which are three-dimensional polymeric structures capable of water solutions sorption, are highly applied in the field of tissue engineering due to their abilities to mimic natural environments. They also have an ability to react on conditions' changes, such as temperature, pH, or electric field [15]. One of their important parameters is porosity and presence of hydrophilic groups such as amino, hydroxyl, and carboxyl [15], which enables swelling with water and provides space for adhesion and proliferation. Figure 4 presents the results of the swelling-capacity investigations. It can be noticed that all of the prepared samples have excellent swelling capacity.

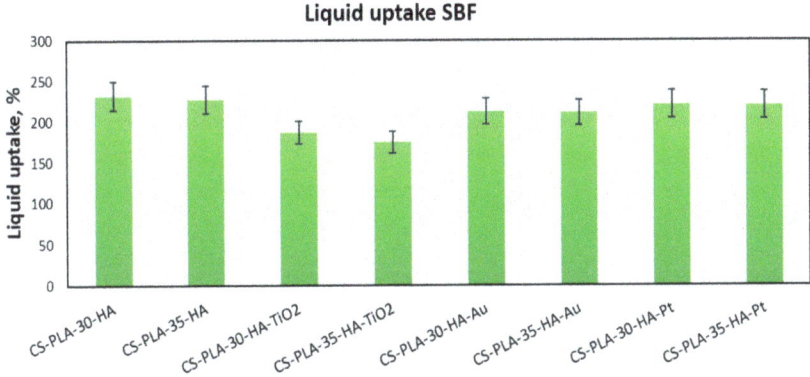

Figure 4. Swelling capacity of the prepared samples.

The best results were obtained for samples CS-PLA-30-HA (232%) and CS-PLA-35-HA (227%), which contained only HA NPs. The scaffolds containing metallic NPs were characterized by slightly worse swelling capacity: 221% and 220% for CS-PLA-30-HA-Pt and CS-PLA-35-HA-Pt, respectively, while for CS-PLA-30-HA-Au, it was 213%, and for CS-PLA-35-HA-Au, it was 211%. The significantly lower sorption efficiency was obtained for the samples containing TiO_2 nanoparticles, and this can be caused by pores' clogging by agglomerated NPs which hampered water molecules' migration inside the

3D matrix. What is important is that the studies were performed in the simulated body solution, which contained various ions. Their presence in many cases decreases sorption abilities of the hydrogels due to their interactions with functional groups. In the case of proposed supporting tissue-regeneration materials, all of them exhibited SC above 200%, meaning that they are suitable for bone recovery [1,5,6]. High sorption ability is especially important during the biomineralization process, which significantly hampers water absorption by hydrogels [41]. An excellent swelling capacity also gives them the possibility of incorporation growth factors or genes. Additionally, swelling degree in terms of a scaffold volume change after liquid uptake was investigated (Figure 5). One may observe that all samples, after 5 min, changed their volume, as a result of the soaking up with simulated body fluid. However, the increase was not very high, and this can be explained by the fact that the pores, after contact with water molecules, decreased due to the hydrophilic interactions with free hydroxyl, carboxyl, and amino groups of the scaffold and electrostatic interactions. The swelling degree is correlated with the swelling capacity. Notably, a small change (decrease) of the scaffolds' volume was observed in time, and the volume stabilized after one hour. Taking everything under consideration, although a small change in the samples' dimensions was observed, it can be omitted due to the fact that no post-soaking volume increase was observed. The scaffolds are immersed in aquatic solutions (SBF, phosphate buffer, etc.) before implementation inside a patient's body. If it does not increase its volume in time, it may be assumed that it will not cause any damage to the tissues and nerves placed nearby [15].

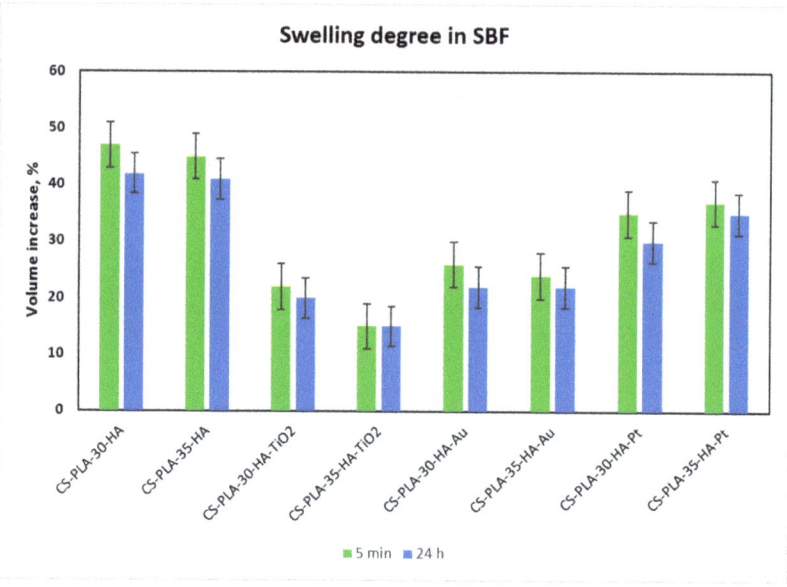

Figure 5. Swelling degree of the prepared samples.

3.3. XRD Analysis

The scaffolds were prepared by using both organic and inorganic components. Apart from chitosan and PLA, they contained inorganic phase, namely hydroxyapatite. The biomaterials were additionally functionalized by other nanoparticles types, such as titanium dioxide, gold, and platinum. To confirm the crystallinity with ICDD 9–432 of the inorganic components, XRD analysis was carried out for the obtained compounds. The results presented in Figure 6a prove preparation of crystalline hydroxyapatite, whereas Figure 6b shows that prepared metal oxide is TiO_2 nanoparticles.

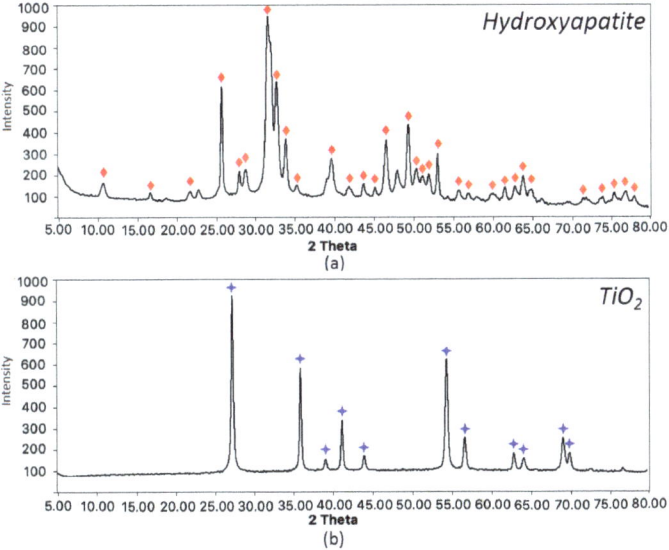

Figure 6. XRD analysis of the scaffolds components: (**a**) hydroxyapatite—the matching patterns are marked with red diamond; (**b**) titanium dioxide—the matching patterns are marked with blue stars.

3.4. Morphology Study of the 3D Scaffolds, and Its Components

To determine morphological properties of the scaffolds, as well as their semi-components, their structure was analyzed by SEM and TEM microscopy. Figure 7a shows pure crosslinked chitosan aerogel, which has a highly porous structure, with pores dimension in the range between 100 and 500 µm, thus providing appropriate conditions for osteoblasts growth, proliferation, and osteogenesis process meeting requirements for bone-tissue engineering [1,5,6]. The scaffold was further modified with synthetic hydroxyapatite, to increase its affinity to this type of tissue and bioactivity (Figure 7b). It can be noticed that HA is well-dispersed on the scaffold surface and does not block the pores. The HA coverage is uniform. Figure 7c shows XRF analysis of the scaffold elemental composition. It can be noticed that the Ca/P ratio is typical for HA (1.6 approximately) and no contaminants such as heavy metal ions are present [40]. The chitosan scaffolds were further modified with PLA nanofibers, to improve its surface properties [2–5,40]. Figure 7d,e presents the biomaterials surface with incorporated HA nanoneedles, which have a length below 100 nm and width around 20 nm (Figure 7f). The average diameter is 72 nm (length), while the polydispersity index is 0.31. The PLA nanofibers are homogenous and continue. They have coaxial morphology, which means that the fibers have a core–shell structure. The nanofibers obtained under 30 kV are slightly thicker than those prepared under 35 kV. Their density is low enough to provide free diffusion of the gases and nutrients. Their surface is regular. The fiber dimensions are in the range of 0.61–1.56 µm. The average diameter is below 1 µm [2–4]. It can be noticed that the nanohydroxyapatite is well dispersed inside the polymeric matrix, in both cases, and are visible in deeper layers. The scaffolds were also functionalized with other nanoparticles, to improve their bioactivity. Figure 7g,h shows biomaterials doped with TiO_2 NPs, which are known for their biosafety and applicability in bone-tissue regeneration [32–35]. They have semi-conductive properties. The titanium dioxide particles are present on the nanofibers surface. However, the SEM images show the spot coverage, not a uniform one, and the NPs are partially aggregated due to their hydrophilic nature in contrast to PLA [11]. The nanoparticles are visible only at the top layer of the biomaterial. There are no significant differences between CS-PLA-30-HA-TiO_2 and CS-PLA-35-HA-TiO_2 samples. TEM photographs (Figure 7i) present TiO_2 nanoparticles that seem to have an amorphous structure, round shape, and size below 100 nm. The average diameter is 95 nm, while the PDI is 0.18.

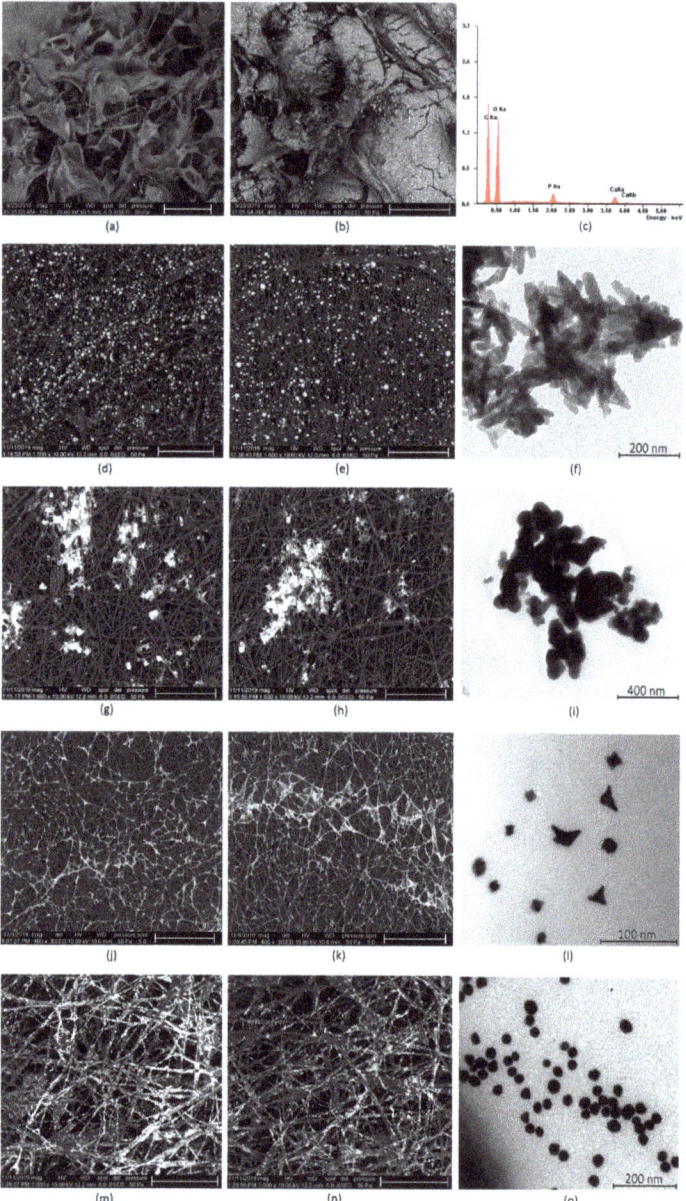

Figure 7. (**a**) SEM microphotograph of the pure chitosan scaffold (scale bar 500 µm); (**b**) SEM microphotograph of the chitosan scaffolds modified with HA nanoparticles (scale bar 200 µm); (**c**) elemental analysis of the scaffold composition; (**d**) SEM microphotograph of the PLA nanofibers obtained under 30kV covered with HA particles (scale bar 50 µm); (**e**) SEM microphotograph of the PLA nanofibers obtained under 35kV covered with HA particles (scale bar 50 µm); (**f**) TEM microphotograph

of the prepared HA nanoparticles; (**g**) SEM microphotograph of the PLA nanofibers obtained under 30 kV covered with TiO_2 nanoparticles (scale bar 50 µm); (**h**) SEM microphotograph of the PLA nanofibers obtained under 35 kV covered with TiO_2 nanoparticles (scale bar 50 µm); (**i**) TEM microphotograph of the prepared TiO_2 nanoparticles; (**j**) SEM microphotograph of the PLA nanofibers obtained under 30 kV covered with Pt nanoparticles (scale bar 50 µm); (**k**) SEM microphotograph of the PLA nanofibers obtained under 35 kV covered with Pt nanoparticles (scale bar 50 µm); (**l**) TEM microphotograph of the prepared Pt nanoparticles; (**m**) SEM microphotograph of the PLA nanofibers obtained under 30 kV covered with Au nanoparticles (scale bar 30 µm); (**n**) SEM microphotograph of the PLA nanofibers obtained under 35 kV covered with Au nanoparticles (scale bar 30 µm); (**o**) TEM microphotograph of the Au nanoparticles.

Next, samples were modified with metallic conductive particles (gold and platinum). Figure 7j,k shows CS-PLA-30-HA-Pt and CS-PLA-35-HA-Pt samples doped with Pt nanoparticles. It can be noticed that the NPs are excellently dispersed and exhibit high affinity to PLA fibers. Interconnected fibers covered with conductive platinum particles may provide good current flow. Again, the NPs are present on the superficial fibers. The morphology of the modified fibers (30 and 35 kV) is almost identical. Interestingly, the Pt NPs created a "spiderweb-like" structure. Figure 7l shows platinum nanoparticles. It can be observed that they are of the size between 10 and 20 nm and of various typical for Pt shapes, like rounded, squared, and triangular. The average diameter is 18 nm, while the PDI is 0.28. Finally, Figure 7m,n provides microphotograph of the nanofibers modified with gold nanoparticles. The Au NPs are well dispersed in the polymeric matrix and exhibit very good affinity to poly (lactic acid) fibers. Most of the superficial fibers are almost fully covered with metallic NPs. What is important, on the contrary to both TiO_2 and Pt NPs, these nanoparticles are present both at the surface of the fibrous layer, as well as in the deeper situated ones. However, in the case of CS-PLA-30-HA-Au sample, the Au NPs are mostly visible at external fibers on the opposite to CS-PLA-35-HA-Au, where gold nanoparticles are located more uniformly. This can be explained by the difference in the fibers' diameter compared to PLA NFs prepared under 30 kV (wider fibers) and 35 kV (narrower fibers). Such NPs arrangement should provide good current flow and fiber conductivity. Figure 7o shows gold nanoparticles size and morphology. The particles are of a round, uniform shape, 20–40 nm. The average diameter is 35 nm, while the PDI is 0.14. All of the prepared NPs exhibit typical morphology and should not exhibit cell toxicity during direct contact, as they cannot penetrate the cell membrane nor damage it, due to their lack of sharp edges [25–27]. Their incorporation should provide extraordinary properties to the hierarchized scaffolds and promote biomineralization process [27,32–35]. The proposed structure should enable cell adhesion and proliferation, along with osteogenic differentiation, since they meet architectural requirements for bone-tissue regeneration [40,42–46].

3.5. Biomineralization Study

The biomineralization process is crucial during bone-tissue regeneration. Thus, scaffolds should not only stimulate osteoblasts' proliferation but also apatite formation. Mineralization may be induced by the presence of functional groups, such as amino, carboxyl, and hydroxyl, which can bind Ca and P cations [40–42]. The results of biomineralization carried out for seven days, under simulated in vitro conditions with and without DC stimulation, are given in Table 2. After one week of bioactivity study, in all samples, mineral sediment was observed. To determine differences between samples, CS-PLA-30-HA (100%) was used as a reference. Standard biomineralization process, which was carried out in simulated body fluid, showed that the addition of TiO_2 nanoparticles positively affects HA formation, imitating the in vivo process, since the biomineralization of CS-PLA-30-HA-TiO_2 is higher by 17% compared to CS-PLA-30-HA and by 18% in the case of CS-PLA-35-HA-TiO_2 compared to CS-PLA-35-HA. In the case of other samples, no superior bioactivity was achieved. There are various methods applied so to accelerate biomineralization, like increased SBF concentration, agitation, or temperature [40]. In this study, an alternative method to enhance biomineralization was proposed based on DC stimulation. The results given in Table 2 show that the application of direct current causes improvement of biomineralization in the case of samples doped with conductive metallic nanoparticles (Pt and Au), while in the case of samples functionalized with TiO_2 NPs (semiconductor) and HA, no effect was observed. Interestingly, the highest increased of the biomineralization occurred for CS-PLA-30-HA-Au (122%) and CS-PLA-35-HA-Au samples (124%). This can be attributed to the highest conductivity of the gold NPs, as well as the highest coverage of the PLA nanofibers not only on the surface but also on the deeper situated fibers. The improvement in the apatite formation can be explained by the forced ions migration present in SBF (calcium, phosphorous) due to the current flow. Moreover, Au NPs are very well-dispersed, which can lead to the increased number of crystallization nuclei. The fibrous/porous membrane-like structure of the scaffolds causes electroosmosis process, which additionally affects the mineralization process. The results obtained for CS-PLA-30-HA-Pt (116%) and CS-PLA-35-HA-Pt (119%) are slightly worse, and they can be attributed to the lower conductivity of the Pt NPs, as well as their presence only on the external fibers. Finally, it may be noticed that, almost in all cases, the samples prepared using PLA nanofibers obtained under 35 kV, exhibiting better properties in terms of biomineralization than those prepared under 30 kV, which was up to 5% for CS-PLA-30-HA and CS-PLA-35-HA samples.

Table 2. Biomineralization efficiency after seven days of incubation in SBF.

Sample	Standard Biomineralization, %	DC-Induced Biomineralization, %	Ca/P Ratio, -
CS-PLA-30-HA	100	98	1.67
CS-PLA-35-HA	105	99	1.66
CS-PLA-30-HA-TiO_2	117	99	1.66
CS-PLA-35-HA-TiO_2	118	99	1.67
CS-PLA-30-HA-Au	100	122	1.70
CS-PLA-35-HA-Au	99	124	1.71
CS-PLA-30-HA-Pt	98	116	1.69
CS-PLA-35-HA-Pt	98	119	1.70

In both cases, the main process involves nucleation, crystallization, and finally crystals' growth, which corresponds to in vivo mineralization [40]. The proposed mechanism of the biomineralization on the scaffolds' surface involves the presence of hydroxyl and carboxyl groups at the mineralization sites coming from N-grafted aspartic, as well as degraded chitosan mers inducing apatite nucleation. The first stage occurs due to the calcium ions binding by anionic groups present on the scaffolds, a finding which corresponds to other researchers' data [36,40–42].

Figure 8 presents the scaffold surface after seven days of in vitro biomineralization. Sedimented ions, which formed mineral precipitate, can be noticed. The coating is quite smooth and uniform, as is desired [40]. XRF analysis confirmed hydroxyapatite formation.

3.6. Cytotoxicity Study of the Prepared Scaffolds

The newly developed scaffolds are dedicated to bone-tissue engineering, so one of the principal properties for these biomaterials is biocompatibility. The cytotoxicity study was carried out on an MG-63 human osteosarcoma cell line, which is typically used for this purpose [39]. Figure 9 shows results of cell culture in the presence of scaffolds. After seven days of culture, MG-63 cells' growth was not disturbed. It can be noticed that, depending on the chemical composition, the number of viable cells is different. Although in all cases the % of living cells is above 100% (control—cells cultured without scaffolds), the presence of various nanoparticles affects osteoblasts' proliferation activity. Two types of biomaterials display significant bioactivity CS-PLA-30-HA-TiO_2 (118%) and CS-PLA-35-HA-TiO_2 (117%). This result corresponds with other researchers' data since titanium dioxide nanomaterials are known for their biosafety with bone-tissue cells [32–35]. Slightly worse results were obtained for CS-PLA-30-HA (112%) and CS-PLA-35-HA (113%), and this is not surprising, since the presence of the hydroxyapatite nanoparticles is known to have a positive effect on bone cells [9,27]. Notably, the scaffolds doped with metallic nanoparticles also exhibited good biocompatibility. Gold NPs are well-studied nanomaterials; however, it is known that their in vitro behavior depends on particle size and shape, as well as the presence of stabilizing agents [25–27]. Au NPs prepared for this study due to the size above 10 nm do not penetrate nor damage cell membrane and do not affect cell cycles, a finding which corresponds to other researchers' results [27,29,30]. The results show that, after seven days, the cell number for sample CS-PLA-30-HA-Au was 110% and CS-PLA-35-HA-Au was 108%; thus, their addition did not cause cytotoxic effect. Finally, samples modified with Pt nanoparticles showed 105% and 102% of cells comparing to the reference. Lack of cytotoxicity can be assigned to their size higher than 10 nm and lack of sharp edges. Additionally, in contrast to silver nanoparticles [28], they are characterized by very good chemical stability and are known to disarm reactive oxygen species (ROS), which may have a positive effect on cells' proliferation, since they prevent their apoptosis triggered by oxygen radicals [25–27,31]. Lack of scaffolds' toxicity can be also assigned to the well-known biosafety of semi-components, such as chitosan [34–36] and PLA [2–7], which are widely used for bone-tissue regeneration [38]. Moreover, the scaffolds' architecture seems to provide good conditions for cells' growth and do not hamper their natural behaviors [38]. The findings show that all of the prepared scaffolds can be considered to be non-toxic and can be used for further studies, which should be focused on in vivo experiments [42–46].

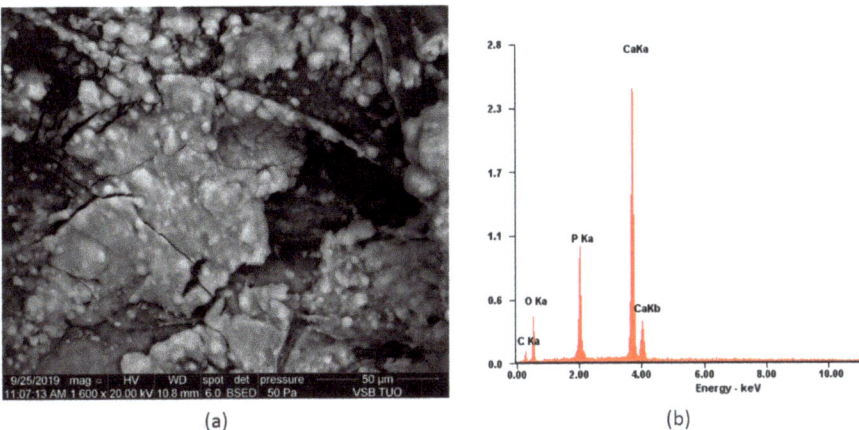

Figure 8. Biomineralization process on the scaffold CS-PLA-30-HA-Au (a) SEM microphotograph of the sample; (b) elemental composition of the sample.

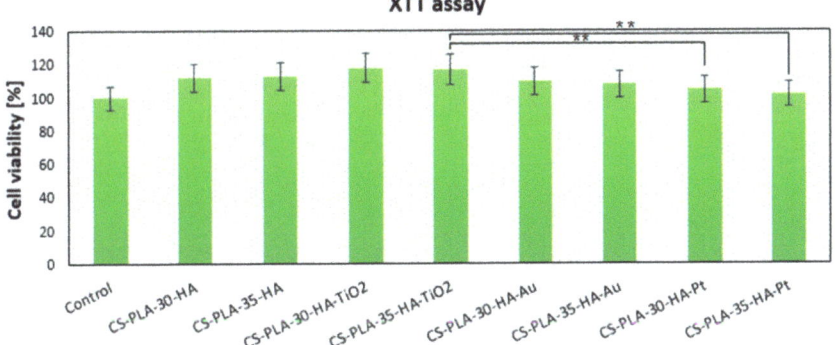

Figure 9. Metabolic activity of osteosarcoma MG-63 cells after seven days of in vitro cell culture determined by XTT assay, (** $p < 0.01$).

4. Conclusions

The aim of the following study was to develop novel 3D bioactive scaffolds with hierarchical structure, using a combination of crosslinked chitosan, electrospinning, and conductive nanoparticles. The innovative scaffolds exhibited extraordinary properties, such as high porosity, excellent swelling properties, and ability of biomineralization electrostimulation. Moreover, the new biomaterials met the requirements for bone-tissue engineering and had a positive impact on MG-63 cells' proliferation activity. The findings showed that the highest bioactivity in contact with cells exhibited samples modified with nanohydroxyapatite and amorphous titanium dioxide NPs, while scaffolds containing nanogold showed highest positive impact on DC-stimulated in vitro biomineralization. Owing to nanostructured architecture, physicochemical properties, and biocompatibility, the proposed scaffold may play an important role in bone-tissue engineering development.

Author Contributions: Conceptualization, M.P., J.R.-P., and Ł.J.; methodology, M.P.; investigation, D.M., M.P., J.R.-P., and Ł.J.; resources, D.M.; M.P., and Ł.J; writing—original draft preparation, J.R.-P. and Ł.J.; supervision, M.P. and D.B.; project administration, M.P.; funding acquisition, M.P. All authors have read and agreed to the published version of the manuscript.

Funding: This research was funded by The National Centre for Science and Development, Poland, grant number [LIDER/42/0149/L-9/17/NCBR/2018]; The Foundation for Polish Science, grant number [START 073.2019].

Conflicts of Interest: The authors declare no conflict of interest.

References

1. Turnbull, G.; Clarke, J.; Picard, F.; Riches, P.; Jia, L.; Han, F.; Li, B.; Shu, W. 3D bioactive composite scaffolds for bone tissue engineering. *Bioact. Mater.* **2018**, *3*, 278–314. [CrossRef] [PubMed]
2. Bhattarai, D.P.; Aguilar, L.E.; Park, C.H.; Kim, C.S. A Review on Properties of Natural and Synthetic Based Electrospun Fibrous Materials for Bone Tissue Engineering. *Membranes* **2018**, *8*, 62. [CrossRef] [PubMed]
3. Di Martino, A.; Liverani, L.; Rainer, A.; Salvatore, G.; Trombetta, M.; Denaro, V. Electrospun scaffolds for bone tissue engineering. *Musculoskelet. Surg.* **2011**, *95*, 69–80. [CrossRef] [PubMed]
4. Khajavi, R.; Abbasipour, M.; Bahador, A. Electrospun biodegradable nanofibers scaffolds for bone tissue engineering. *J. Appl. Polym. Sci.* **2016**, *133*, 42883. [CrossRef]
5. Qu, H.; Fu, H.; Hana, Z.; Sun, Y. Biomaterials for bone tissue engineering scaffolds: A review. *RSC Adv.* **2019**, *9*, 26252–26262. [CrossRef]
6. Aslankoohi, N.; Mondal, D.; Rizkalla, A.S.; Mequanint, K. Bone Repair and Regenerative Biomaterials: Towards Recapitulating the Microenvironment. *Polymers* **2019**, *11*, 1437. [CrossRef]
7. Permyakova, E.S.; Kiryukhantsev-Korneev, P.V.; Gudz, K.Y.; Konopatsky, A.S.; Polčak, J.; Zhitnyak, I.Y.; Gloushankova, N.A.; Shtansky, D.V.; Manakhov, A.M. Comparison of Different Approaches to Surface Functionalization of Biodegradable Polycaprolactone Scaffolds. *Nanomaterials* **2019**, *9*, 1769. [CrossRef]
8. Witzler, M.; Büchner, D.; Shoushrah, S.H.; Babczyk, P.; Baranova, J.; Witzleben, S.; Tobiasch, E.; Schulze, M. Polysaccharide-Based Systems for Targeted Stem Cell Differentiation and Bone Regeneration. *Biomolecules* **2019**, *9*, 840. [CrossRef]
9. Fang, C.-H.; Lin, Y.-W.; Lin, F.-H.; Sun, J.-S.; Chao, Y.-H.; Lin, H.-Y.; Chang, Z.-C. Biomimetic Synthesis of Nanocrystalline Hydroxyapatite Composites: Therapeutic Potential and Effects on Bone Regeneration. *Int. J. Mol. Sci.* **2019**, *20*, 6002. [CrossRef]
10. Xu, Z.; Wang, N.; Liu, P.; Sun, Y.; Wang, Y.; Fei, F.; Zhang, S.; Zheng, J.; Han, B. Poly(Dopamine) Coating on 3D-Printed Poly-Lactic-Co-Glycolic Acid/β-Tricalcium Phosphate Scaffolds for Bone Tissue Engineering. *Molecules* **2019**, *24*, 4397. [CrossRef]
11. Narayanan, G.; Vernekar, N.; Kuyinu, E.L.; Laurencin, C.T. Poly (Lactic Acid)-Based Biomaterials for Orthopaedic Regenerative Engineering. *Adv. Drug Deliv. Rev.* **2016**, *107*, 247–276. [CrossRef] [PubMed]
12. Rogina, A. Electrospinning process: Versatile preparation method for biodegradable and natural polymers and biocomposite systems applied in tissue engineering and drug delivery. *Appl. Surf. Sci.* **2014**, *296*, 221–230. [CrossRef]
13. Bhattarai, R.S.; Bachu, R.D.; Boddu, S.H.S.; Bhaduri, S. Biomedical Applications of Electrospun Nanofibers: Drug and Nanoparticle Delivery. *Pharmaceutics* **2019**, *11*, 5. [CrossRef] [PubMed]

14. Cassan, D.; Becker, A.; Glasmacher, B.; Roger, Y.; Hoffmann, A.; Gengenbach, T.R.; Easton, C.D.; Hänsch, R.; Menzel, H. Blending chitosan-g-poly(caprolactone) with poly(caprolactone) by electrospinning to produce functional fiber mats for tissue engineering applications. *J. Appl. Polym. Sci.* **2019**, *94*, 48650. [CrossRef]
15. Yusof, M.R.; Shamsudin, R.; Zakaria, S.; Hamid, M.A.A.; Yalcinkaya, F.; Abdullah, Y.; Yacob, N. Fabrication and Characterization of Carboxymethyl Starch/Poly(l-Lactide) Acid/β-Tricalcium Phosphate Composite Nanofibers via Electrospinning. *Polymers* **2019**, *11*, 1468. [CrossRef]
16. Chan, K.V.; Asadian, M.; Onyshchenko, I.; Declercq, H.; Morent, R.; De Geyter, N. Biocompatibility of Cyclopropylamine-Based Plasma Polymers Deposited at Sub-Atmospheric Pressure on Poly (ε-caprolactone) Nanofiber Meshes. *Nanomaterials* **2019**, *9*, 1215. [CrossRef]
17. Braghirolli, D.I.; Steffens, D.; Pranke, P. Electrospinning for regenerative medicine: a review of the main topics. *Drug Discov Today.* **2014**, *19*, 743–753. [CrossRef]
18. Niiyama, E.; Uto, K.; Lee, C.M.; Sakura, K.; Ebara, M. Alternating Magnetic Field-Triggered Switchable Nanofiber Mesh for Cancer Thermo-Chemotherapy. *Polymers* **2018**, *10*, 1018. [CrossRef]
19. Szentivanyi, A.L.; Zernetsch, H.; Menzel, H.; Glasmacher, B. A review of developments in electrospinning technology: New opportunities for the design of artificial tissue structures. *Int. J. Artif. Organs* **2011**, *34*, 986–997. [CrossRef]
20. Shahidi, F.; Abuzaytoun, R. Chitin, chitosan, and co-products: Chemistry, production, applications, and health effects. *Adv. Food Nutr. Res.* **2005**, *49*, 93–135.
21. Dash, M.; Chiellini, F.; Ottenbrite, R.M.; Chiellini, E. Chitosan—A versatile semi-synthetic polymer in biomedical applications. *Prog. Polym. Sci.* **2011**, *36*, 981–1014. [CrossRef]
22. Kean, T.; Thanou, M. Biodegradation, biodistribution and toxicity of chitosan. *Adv. Drug Deliv. Rev.* **2010**, *62*, 3–11. [CrossRef]
23. LogithKumar, R.; KeshavNarayan, A.; Dhivya, S.; Chawla, A.; Saravanan, S.; Selvamurugan, N. A review of chitosan and its derivatives in bone tissue engineering. *Carbohydr. Polym.* **2016**, *151*, 172–188. [CrossRef] [PubMed]
24. Saravanan, S.; Leena, R.S.; Selvamurugan, N. Chitosan based biocomposite scaffolds for bone tissue engineering. *Int. J. Biol. Macromol.* **2016**, *93*, 1354–1365. [CrossRef]
25. Fathi-Achachelouei, M.; Knopf-Marques, H.; Ribeiro da Silva, C.E.; Barthès, J.; Bat, E.; Tezcanerand, A.; Vrana, N.E. Use of Nanoparticles in Tissue Engineering and Regenerative Medicine. *Front. Bioeng. Biotechnol.* **2019**, *7*, 113. [CrossRef]
26. Khan, I.; Saeed, K.; Khan, I. Nanoparticles: Properties, applications and toxicities. *Arab. J. Chem.* **2019**, *12*, 908–931. [CrossRef]
27. Vieira, S.; Vial, S.; Reis, R.L.; Oliveira, J.M. Nanoparticles for bone tissue engineering. *Biotechnol. Prog.* **2017**, *33*, 590–611. [CrossRef] [PubMed]
28. AshaRani, P.V.; Low Kah Mun, G.; Hande, M.P.; Valiyaveettil, S. Cytotoxicity and genotoxicity of silver nanoparticles in human cells. *ACS Nano* **2009**, *3*, 279–290. [CrossRef] [PubMed]
29. Chithrani, B.D.; Ghazani, A.A.; Chan, W.C. Determining the size and shape dependence of gold nanoparticle uptake into mammalian cells. *Nano Lett.* **2006**, *6*, 662–668. [CrossRef]
30. Chompoosor, A.; Saha, K.; Ghosh, P.S.; Macarthy, D.J.; Miranda, O.R.; Zhu, Z.J.; Arcaro, K.F.; Rotello, V.M. The role of surface functionality on acute cytotoxicity, ROS generation and DNA damage by cationic gold nanoparticles. *Small* **2010**, *6*, 2246–2249. [CrossRef]
31. Pedone, D.; Moglianetti, M.; De Luca, E.; Giuseppe, B.; Pompa, P.P. Platinum nanoparticles in nanobiomedicine. *Chem. Soc. Rev.* **2017**, *46*, 4951–4975. [CrossRef] [PubMed]
32. Brammer, K.S.; Frandsen, C.J.; Jin, S. TiO_2 nanotubes for bone regeneration. *Trends Biotechnol.* **2012**, *30*, 315–322. [CrossRef] [PubMed]
33. Lee, J.-E.; Bark, C.W.; Quy, H.V.; Seo, S.-J.; Lim, J.-H.; Kang, S.-A.; Lee, Y.; Lee, J.-M.; Suh, J.-Y.; Kim, Y.-G. Effects of Enhanced Hydrophilic Titanium Dioxide-Coated Hydroxyapatite on Bone Regeneration in Rabbit Calvarial Defects. *Int. J. Mol. Sci.* **2018**, *19*, 3640. [CrossRef] [PubMed]
34. Kumar, P. Nano-TiO_2 Doped Chitosan Scaffold for the Bone Tissue Engineering Applications. *Int. J. Biomater.* **2018**, *2018*, 6576157. [CrossRef] [PubMed]
35. Ikono, R.; Li, N.; Pratama, N.H.; Vibriani, A.; Yuniarni, D.R.; Luthfansyah, M.; Bachtiar, B.M.; Bachtiar, E.W.; Mulia, K.; Nasikin, M.; et al. Enhanced bone regeneration capability of chitosan sponge coated with TiO_2 nanoparticles. *Biotechnol. Rep.* **2019**, *24*, e00350. [CrossRef]

36. Guo, N.; Zhang, L.; Wang, J.; Wang, S.; Zoud, Y.; Wang, X. Novel fabrication of morphology tailored nanostructures with Gelatin/Chitosan Co-polymeric bio-composited hydrogel system to accelerate bone fracture healing and hard tissue nursing care management. *Process. Biochem.* **2020**, *90*, 177–183. [CrossRef]
37. Santoro, M.; Shah, S.R.; Walker, J.L.; Mikos, A.G. Poly(lactic acid) nanofibrous scaffolds for tissue engineering. *Adv. Drug Deliv. Rev.* **2016**, *107*, 206–212. [CrossRef]
38. Coltelli, M.B.; Cinelli, P.; Gigante, V.; Aliotta, L.; Morganti, P.; Panariello, L.; Lazzeri, A. Chitin Nanofibrils in Poly(Lactic Acid) (PLA) Nanocomposites: Dispersion and Thermo-Mechanical Properties. *Int. J. Mol. Sci.* **2019**, *20*, 504. [CrossRef]
39. Pautke, C.; Schieker, M.; Tischer, T.; Kolk, A.; Neth, P.; Mutschler, W.; Milz, S. Characterization of osteosarcoma cell lines MG-63, Saos-2 and U-2 OS in comparison to human osteoblasts. *Anticancer Res.* **2004**, *24*, 3743–3748.
40. Katsanevakis, E.; Wen, X.J.; Shi, D.L.; Zhang, N. Biomineralization of Polymer Scaffolds. *Key Eng. Mater.* **2010**, *441*, 269–295. [CrossRef]
41. Lu, H.T.; Lu, T.W.; Chen, C.H.; Lu, K.Y.; Mi, F.L. Development of nanocomposite scaffolds based on biomineralization of N,O-carboxymethyl chitosan/fucoidan conjugates for bone tissue engineering. *Int. J. Biol. Macromol.* **2018**, *120*, 2335–2345. [CrossRef] [PubMed]
42. Feng, P.; Kong, Y.; Yu, L.; Li, Y.; de Gao, C.; Peng, S.; Pan, H.; Zhao, Z.; Shuai, C. Molybdenum disulfide nanosheets embedded with nanodiamond particles: Co-dispersion nanostructures as reinforcements for polymer scaffolds. *Appl. Mater. Today* **2019**, *17*, 216–226. [CrossRef]
43. Shuai, C.; Yang, W.; He, C.; Peng, S.; Gao, C.; Yang, Y.; Qi, F.; Feng, P. A magnetic micro-environment in scaffolds for stimulating bone regeneration. *Mater. Des.* **2020**, *185*, 108275. [CrossRef]
44. Shuai, C.; Liu, G.; Yang, Y.; Yang, W.; He, C.; Wang, G.; Liu, Z.; Qi, F.; Peng, S. Functionalized BaTiO3 enhances piezoelectric effect towards cell response of bone scaffold. *Colloids Surf. B Biointerfaces* **2020**, *185*, 110587. [CrossRef]
45. Feng, P.; Wu, P.; Gao, C.; Yang, Y.; Guo, W.; Yang, W.; Shuai, C. A Multimaterial Scaffold With Tunable Properties: Toward Bone Tissue Repair. *Adv. Sci.* **2018**, *5*, 1700817. [CrossRef]
46. Shuai, C.; Yu, L.; Yang, W.; Peng, S.; Zhong, Y.; Feng, P. Phosphonic Acid Coupling Agent Modification of HAP Nanoparticles: Interfacial Effects in PLLA/HAP Bone Scaffold. *Polymers* **2020**, *12*, 199. [CrossRef]

© 2020 by the authors. Licensee MDPI, Basel, Switzerland. This article is an open access article distributed under the terms and conditions of the Creative Commons Attribution (CC BY) license (http://creativecommons.org/licenses/by/4.0/).

Article

3D Ultrasensitive Polymers-Plasmonic Hybrid Flexible Platform for In-Situ Detection

Meimei Wu [1], Chao Zhang [1,2], Yihan Ji [1], Yuan Tian [1], Haonan Wei [1], Chonghui Li [1], Zhen Li [1], Tiying Zhu [1], Qianqian Sun [1], Baoyuan Man [1,2] and Mei Liu [1,2,*]

1. School of Physics and Electronics, Shandong Normal University, Jinan 250014, China; sdnumm@163.com (M.W.); czsdnu@126.com (C.Z.); Ji_Yi_Han@163.com (Y.J.); 2018020524@stu.sdnu.edu.cn (Y.T.); 2018020530@stu.sdnu.edu.cn (H.W.); chonghuili163@163.com (C.L.); lizhen19910528@163.com (Z.L.); 2018020534@stu.sdnu.edu.cn (T.Z.); qianqiansun@sdnu.edu.cn (Q.S.); byman@sdnu.edu.cn (B.M.)
2. Institute of Materials and Clean Energy, Shandong Normal University, Jinan 250014, China
* Correspondence: liumei@sdnu.edu.cn

Received: 3 January 2020; Accepted: 30 January 2020; Published: 9 February 2020

Abstract: This paper introduces a three-dimensional (3D) pyramid to the polymers-plasmonic hybrid structure of polymethyl methacrylate (PMMA) composite silver nanoparticle (AgNPs) as a higher quality flexible surface-enhanced Raman scattering (SERS) substrate. Benefiting from the effective oscillation of light inside the pyramid valley could provide wide distributions of 3D "hot spots" in a large space. The inclined surface design of the pyramid structure could facilitate the aggregation of probe molecules, which achieves highly sensitive detection of rhodamine 6G (R6G) and crystal violet (CV). In addition, the AgNPs and PMMA composite structures provide uniform space distribution for analyte detection in a designated hot spot zone. The incident light can penetrate the external PMMA film to trigger the localized plasmon resonance of the encapsulated AgNPs, achieving enormous enhancement factor (~6.24×10^8). After undergoes mechanical deformation, the flexible SERS substrate still maintains high mechanical stability, which was proved by experiment and theory. For practical applications, the prepared flexible SERS substrate is adapted to the in-situ Raman detection of adenosine aqueous solution and the methylene-blue (MB) molecule detection of the skin of a fish, providing a direct and nondestructive active-platform for the detecting on the surfaces with any arbitrary morphology and aqueous solution.

Keywords: SERS; PMMA; AgNPs; in-situ; adenosine; methylene-blue

1. Introduction

With ultra-sensitive, rapid, and non-destructive properties, surface-enhanced Raman scattering (SERS) is generally considered a valuable analytical technique for achieving label-free detection of biochemical molecules [1–3]. Recent research concerning SERS active substrates focused on designing and fabricating metallic nanostructures with tunable plasmonic features on hard substrates [4–7]. Significant Raman enhancement tends to occur at nanogaps between neighboring metal nanostructures with high local-field intensities (so-called "hot spots") [8–10]. However, in many cases, SERS substrates based on silicon wafers or glass slides are rigid, brittle, and hence unsuitable for directly analyzing target molecules in a solution or gas [11,12]. Compared to traditional rigid substrates, flexible substrates can wrap around complex surfaces and be tailored to the desired size and shape [13]. In addition to achieving a highly active SERS substrate, the effective and rapid collection of probe molecules to the substrate presents a significant challenge [12].

Generally, flexible substrates with simple and periodic two-dimensional (2D) structures are limited by their low density of hot spots distributed at the gaps between horizontal nanoparticals

(NPs), which leads to weak Raman enhancement [14]. Thus, to further improve SERS sensitivity, the integration of flexible three-dimensional (3D) structures is imperative. Such 3D flexible SERS substrates can offer large specific surfaces and create high-density hot spots to capture more target molecules [15]. Recently, great efforts were dedicated to fabricating highly sensitive flexible substrates by combining noble plasmonic nanoparticles with polymers. Park et al. have demonstrated a transparent and flexible SERS substrate based on the polydimethylsiloxane (PDMS) film embedded with gold nanostar (GNS) [13]. Kumar et al. fabricated a 3D flexible SERS substrate by depositing AgNPs on structured polydimethylsiloxane using a Taro leaf as the template to detect malachite green [16]. Kumar et al. fabricated a buckled PDMS silver nanorods array as an active 3D SERS sensor for detecting bacteria [17]. Despite demonstrating enormous application prospects for real sample detection, these flexible substrates still involve unresolved issues regarding metal-molecule contact and metal particle oxidation, both of which can cause poor reproducibility and stability in SERS detection [17,18].

In this study, we propose a 3D pyramid-shaped AgNPs@PMMA flexible composite platform for SERS application. The inclined surface of the periodic pyramid-shaped structures could facilitate the aggregation of probe molecules and create high-density 3D "hot spots". In addition, the AgNPs and PMMA composite structures provide uniform space distribution for analyte detection in designated hot spot zone, and the incident light can penetrate the external PMMA film to trigger the localized plasmon resonance of the encapsulated AgNPs, achieving enormous enhancement factor ($\sim 6.24 \times 10^8$). The fabricated 3D pyramid-shaped AgNPs@PMMA flexible substrate achieved highly sensitive detection of rhodamine 6G (R6G) and crystal violet (CV), and displayed both high stability and signal reproducibility. Following mechanical stimuli in our experiments, such as stretching and bending, the flexible substrate maintained the stable Raman signal, which corroborates our theoretical simulation. Finally, we successfully achieve the in-situ detection in adenosine aqueous solution and the methylene-blue (MB) molecule detection of the skin of a fish, which demonstrates its great potential for the detecting on the surfaces with any arbitrary morphology and aqueous solution.

2. Materials and Methods

2.1. Materials

Acetone (CH_3COCH_3, 99.5%), alcohol (C_2H_6O, 99.7%), ethylene glycol ($C_2H_6O_2$, 99.0%), R6G, crystal violet (CV), MB, silver nitrate ($AgNO_3$), and polymethyl methacrylate (PMMA) were all obtained from Sinopharm Chemical Reagent Co. Ltd. (Shanghai, China). Polyvinylpyrrolidone (PVP, $Mw = 55,000$) was obtained from Sigma-Aldrich (St. Louis, MO, USA), and the adenosine sample was purchased from Sangon Biotech (Shanghai) Co. Ltd. (Shanghai, China). The adenosine was dissolved in ultrapure water to make a 0.01M stock solution and then diluted to the final concentration before use.

2.2. Fabrication of 3D Pyramid AgNPs @PMMA Flexible Substrate

Figure 1 schematically illustrates the fabrication procedure for the 3D-pyramid-shaped AgNPs @PMMA flexible (P–AgNPs@PMMA) substrate. Pyramid Si substrate (P–Si) was pretreated using the method of wet texturing boron-doped monocrystalline silicon in a NaOH solution [19]. AgNPs dispersion was synthesized by the oil bath synthesis, according to previous work [20]. First, 0.075 g of PMMA grains were put into 100 mL of acetone, then this solution was stirred at 80 °C for 1 h. High-density and uniform AgNPs dispersion was added into the PMMA solution, and subsequently, this mixed solution was stirred at room temperature and allowed to stand overnight. The mixed solution was then dip-coated on the P-Si substrate, after the mixed layer was dried, the sample was immersed in a NaOH (30%, 100 mL) solution for 1 h to remove the P–Si supporter. The flexible films were rinsed with running deionized water, inverted on a glass substrate, and dried at room temperature to be used as the SERS substrate in our experiments.

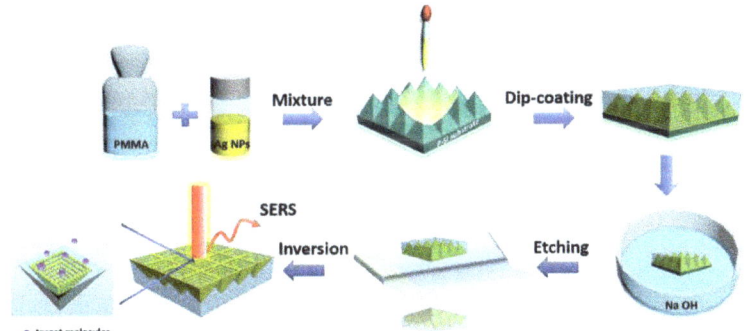

Figure 1. Schematic illustration of the fabrication process for the P–AgNPs@PMMA substrate.

2.3. Characterization

Scanning electron microscopy (SEM) images of the P–AgNPs@PMMA substrate were characterized using a scanning electron microscope (SEM, Sigma 500, Carl Zeiss, Jena, Germany) with an energy dispersive spectrometer (EDS) at a voltage of 10 kV. The SERS spectra of the P–AgNPs@PMMA substrate were recorded by a Raman spectrometer (LabRAM HR Evolution, Horiba, Kyoto, Japan) with 532 nm laser excitation. The effective power of the laser source was kept at 0.048 mW, and a 50× objective (N.A. = 0.50) was used throughout the test. The diffraction grid was 600 gr/nm, and the integration time was 8 s. All the SERS spectra in the experiment are expressed on the basis of average spectra. The UV-visible absorption spectra of the synthesized silver nanoparticles was collected by a dual beam UV-visible spectrophotometer (TU-1900, Beijing General Analysis General Instrument Co., Ltd, Beijing, China).

3. Results and Discussion

To investigate the structural properties of the P–AgNPs@PMMA composite substrates, we performed SEM characterization (Figure 2). AgNPs were deposited on the surface of the P-Si substrate via dip-coating, as shown in Figure 2a. The inclined surface of the periodic pyramid-shaped structures could facilitate the aggregation of probe molecules and create high-density 3D "hot spots", which is critical for the final enhancement activity. Figure 2b,c, respectively exhibit low- and high-magnification SEM images of the P–AgNPs@PMMA flexible substrate after removal of the P-Si substrate. As pictured in Figure S1 (from the Supplementary Materials), the average size of these AgNPs is ~78nm, and the gaps among nanoparticles are very narrow, which supports huge electromagnetic enhancement for absorbed molecules. The composition of the P–AgNPs@PMMA flexible substrate is also analyzed with high-resolution EDS mapping in Figure 2d,e, indicating that high-density AgNPs were successfully embedded within the PMMA film. Figure 2f shows a strong UV-vis absorption peak at ~476 nm of AgNPs, which is attributed to the strong coupling of the plasma between the AgNPs. Thus, a high signal-to-noise ratio SERS signal can be generated at 532 nm incident excitation light.

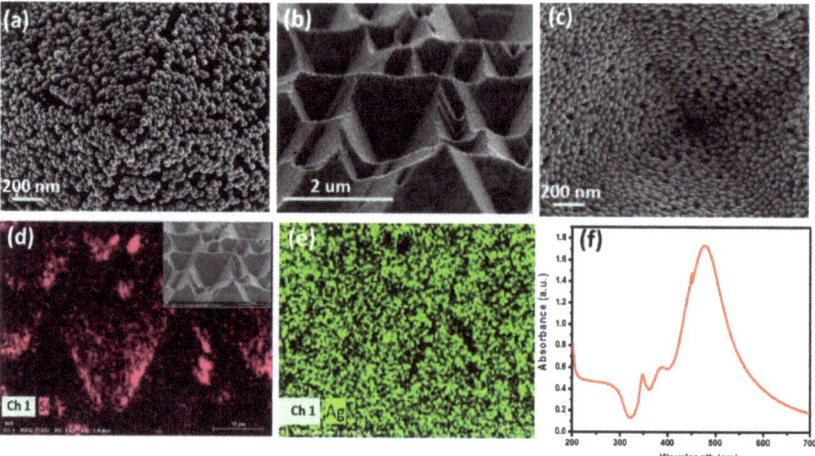

Figure 2. (**a**) Scanning electron microscopy (SEM) image of AgNPs deposited on the P-Si substrate; (**b**) and (**c**) are SEM images of the P–AgNPs@PMMA substrate in different magnification; (**d**) and (**e**) energy dispersive spectrometer (EDS) mapping of the P–AgNPs@ PMMA flexible substrates; (**f**) UV–vis absorption spectra showing the SPR peak of the AgNPs at ~476 nm.

It has been reported that high-density "hot spots" are always generated at the slits or tips of metal nanostructures and the interspacing gaps between the adjacent metal nanostructures could markedly affect the ability of detection limit and the intensity of SERS signals [15]. To further optimize the Raman enhancement effect of the composite SERS substrate, we adjusted the distribution density of AgNPs in the PMMA solution to find the best PMMA/Ag volume ratio. Figure 3a shows the SERS spectra for R6G (10^{-7} M) absorbed on AgNPs@PMMA/P-Si substrates with different PMMA /Ag volume ratio. The characteristic SERS peaks of R6G were observed around 612, 773, 1308, 1358, 1505, and 1645 cm^{-1}. The peak at 612 cm^{-1} is attributed to C–C–C bond stretching modes. The peak at 773 cm^{-1} is due to the R6G C–H out-of-plane bending mode. The peaks around 1308, 1358, 1505, and 1645 cm^{-1} originate from aromatic C–C stretching [21,22]. It is observed that the Raman signal of the R6G molecule shows an apparent trend from the rise to decline with the continuous addition of silver colloid in a volume of PMMA solution. From Figure 3b, we can see that the SERS performance of the substrate is optimal with the PMMA/Ag volume ratio of 2:3. As shown in Figure S2 (from the Supplementary Materials), the SEM images of different volume ratio AgNPs@PMMA/P-Si substrate further illustrate that the number of AgNPs is small and sparse in the initial stage of the reaction. The area where "hot spots" generated between the silver particles is small, and the Raman signal is weak. As silver particles become denser with the continuous addition of silver colloid, the particle spacing decreases, which causes the increases of the density of "hot spots", and further the Raman signal of the molecule gradually increases. When the PMMA/Ag volume ratio was 2:3, the SERS signal of the molecule reached the strongest. As the silver colloid further addition, the AgNPs accumulate to form large clusters, inhibiting the formation of electromagnetic fields between the particles, resulting in a weakening of the Raman signal. Therefore, the AgNPs@PMMA/P-Si substrates with PMMA/Ag volume ratio of 2:3 were determined as optimal.

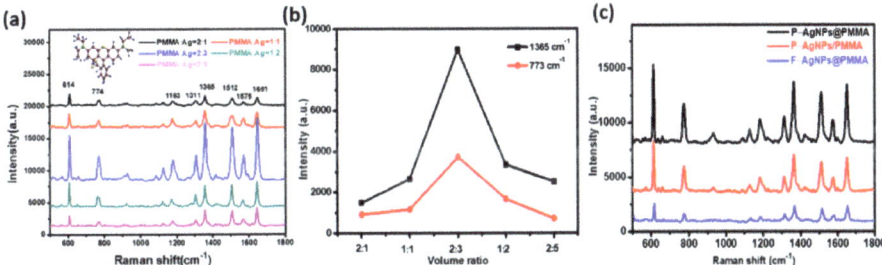

Figure 3. (a) The surface-enhanced Raman scattering (SERS) spectra obtained from the P–AgNPs@PMMA substrate with different PMMA/Ag volume ratio; here, the concentration of rhodamine 6G (R6G) is 10^{-7} M. Insert: the structural formula of R6G molecule; (b) change in Raman intensity of R6G at 774 and 1365 cm^{-1} as a function of volume ratio; (c) SERS spectra of 10^{-7} M R6G absorbed on the P–AgNPs@PMMA (black line), P–AgNPs/PMMA (red line) and F–AgNPs@PMMA (blue line) substrates.

To further investigate the SERS effect of the AgNPs and PMMA composite structures and demonstrate the advantages of this 3D pyramid structure for SERS activity, we analyzed and compared the Raman spectra of R6G (10^{-7} M) molecules adsorbed on pyramid-shaped AgNPs and PMMA composite (P–AgNPs@PMMA), AgNPs attached to the pyramid-shaped PMMA film (P–AgNPs/PMMA) and planar AgNPs and PMMA composite (F–AgNPs@PMMA) flexible SERS substrates in Figure 3c. Compared to F–AgNPs@PMMA, P–AgNPs/PMMA has a higher SERS signal, which indicates that the inclined surface of the 3D pyramid structure could facilitate the aggregation of probe molecules and create high-density 3D "hot spots". In addition, the AgNPs and PMMA composite structures provide uniform space distribution for analyte detection in designated hot spot zone, avoiding the calculation of Raman enhancement factor error caused by different SERS spectra measured in different regions, achieving reliable enhancement factor measurement. The P–AgNPs@PMMA substrate provides a strong enhancement factor equivalent to P–AgNPs/PMMA substrate, and obtains high spatial uniformity and good stable SERS spectrum. The uniform detection geometry can avoid signal distortion caused by direct contact with the probe molecules to cause deformation of the probe molecules, obtaining a more stable SERS signal. As a whole, the high SERS intensity of the composite P–AgNPs@PMMA substrate benefits from the high-density 3D "hot spots". More importantly, the hybrid detection geometry ensures uniform distribution of analyte molecules in the designated hot spot zone, and the incident light can penetrate the external PMMA film to trigger the localized plasmon resonance of the encapsulated AgNPs, achieving strong enhancement factor and stable Raman signal.

R6G and CV, most commonly used for SERS molecules, were chosen to test the SERS performance of the P–AgNPs@PMMA substrate (Figure 4). Figure 4a illustrates the Raman intensities of R6G deposited on the P–AgNPs@PMMA substrate at various molecular concentrations. It can be seen that the intensity of SERS peaks decreased with the decreasing concentration of R6G solution. Obviously, the 614 and 774 cm^{-1} peaks of R6G were still observed on the substrate even when the concentration was decreased to 10^{-13} M, indicating that the substrate exhibits strong Raman effects. Here, we selected the 614 cm^{-1} peak to further investigate the relationship between the Raman intensity and R6G concentration. The reasonable linear response of the P–AgNPs@PMMA substrate is displayed in Figure 4b. The correlation coefficient (R^2) values of the peaks at 614 and 774 cm^{-1} can reach 0.988 and 0.989, respectively. In addition, we have calculated the number of adsorbed molecules at different concentrations. The number of adsorbed molecules are estimated by

$$N = \frac{C_m V_m S_{laser}}{S_S} N_A, \quad (1)$$

where C_m is the concentration of molecular solution, V_m is the volume of molecular aqueous solution dipped onto the whole substrate, S_S is the area of the whole substrate, S_{laser} is the area of the focal spot of laser, and N_A is the Avogadro's constant. In the experiments, 20 µL different concentration R6G and CV aqueous solutions were respectively dipped in the substrate. The diameter of the focused laser spot is ~2 µm. The size of the substrate is 5 mm × 5 mm [23]. We obtained the number of adsorbed molecules in the range of $1.51 \times 10^6 - 0.15$, depending on different molecules concentration (10^{-6}–10^{-13} M). The number of adsorbed molecules gradually increases with the molecular concentration increases. Linear correlation between concentration and intensity can be supported by the number of adsorbed molecules. These results demonstrate the well linear dependence of the substrate. We calculate the enhancement factors (*EF*) for R6G molecule absorbed on the P–AgNPs@PMMA substrate, using the following standard Equation:

$$EF = \frac{I_{SERS}/N_{SERS}}{I_{RS}/N_{RS}}. \tag{2}$$

I_{SERS} and I_{RS} are the peak intensity of the SERS spectra and the Raman intensity obtained from the SERS sample and pure PMMA film, respectively. N_{SERS} and N_{RS} are the numbers of target molecules illuminated by the laser spot on the sample and pure PMMA film respectively [24]. The enhancement factor (**EF**) for 614 cm^{-1} peak of this hybrid structure is calculated as 6.24×10^8. The high EF of the P–AgNPs@PMMA substrate is more superior than the previous metal nanoparticles-based SERS substrates (Table 1), which can be attributed to the combined effect of the 3D-pyramid structure and PMMA film. Pyramid structure demonstrates large specific surfaces, which provide many activity sites, and the inclined surface design of this structure can capture more target molecules. Moreover, the PMMA film acts as a high-dielectric layer and enables the strong surface plasmon coupling between AgNPs embedded in PMMA. The uniform hybrid detection geometry ensures uniform distribution of analyte molecules in the designated hot spot zone, providing reliable enhancement factor measurement, resulting in highly sensitive detection of target molecules.

Table 1. Comparing the performance of various flexible SERS sensors.

SERS Substrate	Analytes	Enhancement Factor (EF)	Ref.
Ag-nanosheet-grafted polyamide-nanofibers	4-mercaptobenzoicacid	2.2×10^7	[25]
Ag decorated microstructured PDMS substrate fabricated from Taro leaf	Malachite green	2.06×10^5	[16]
GNS/PDMS	Benzenedithiol	1.9×10^8	[13]
AuNR/fitter paper	1,4-Benzenedithiol	5×10^6	[26]
Flexible free-standing silver nanoparticle-graphene	Rhodamine6G	1.25×10^7	[27]
AgNP/fitter paper by brushing technique	Rhodamine 6G	2.2×10^7	[28]
Flexible AgNP@PMMA/P-Si	Rhodamine 6G	6.24×10^8	This work

To further investigate SERS acitivity of P–AgNPs@PMMA substrate, we also collected the Raman signal of the CV molecule with different concentrations (Figure 4c). It is well-known that the Raman cross-section of CV molecules is significantly different from R6G at the same excitation wavelength [29]. The molecule's diameter of CV is 15.10 Å [30]. It can be seen that the Raman main peaks located at ~914, 1178, 1533, 1587, and 1620 cm^{-1} are present in the SERS spectra of CV. The two peaks at ~914 and ~1178 cm^{-1} are attributed to the ring skeletal vibration of radical orientation. Meanwhile, the peaks around ~1533, ~1587, and ~1620 cm^{-1} are due to the ring C–C stretching modes. The Raman peaks of CV at 914 cm^{-1} can be observed even though the concentration of CV decreased to 10^{-12} M, reaching the minimum detection limit, which can be attributed to the 3D high-density SERS "hot spot", and the excitation laser can easily penetrate the external PMMA film, causing strong surface plasmon resonance of the embedded AgNPs. Figure 4d shows the Raman intensity of CV changed linearly with the molecule concentration, and the R^2 value of the peaks at 914 and 1587 cm^{-1} can reach 0.990 and 0.993, respectively. We obtained the number of adsorbed molecules in the range of $1.51 \times 10^7 - 1.51$, depending on different molecule concentrations (10^{-5}–10^{-12} M), which present a good linear fit relationship. These results adequately demonstrate that the P–AgNPs@PMMA substrate possesses perfect SERS performance and good quantitative detection capability.

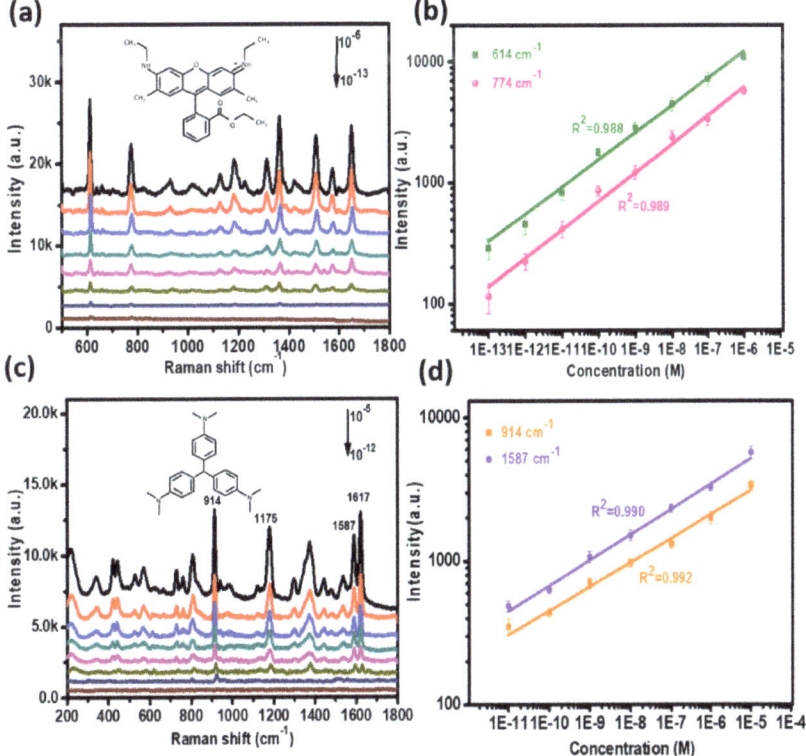

Figure 4. (**a**) Raman spectra of R6G from 10^{-6} to 10^{-13} M on the P–AgNPs@PMMA substrate. Insert: the structural formula of R6G molecule; (**b**) the Raman spectra of R6G at 614 and 774 cm^{-1} as a function of the molecular concentration on the P–AgNPs@PMMA substrate in log scale; (**c**) Raman spectra of CV with concentration from 10^{-5} to 10^{-12} M on the P–AgNPs@PMMA substrate. Insert: the structural formula of CV molecule; (**d**) the Raman spectra of CV at 914 cm^{-1} and 1567 cm^{-1} as a function of the molecular concentration on the P–AgNPs@PMMA substrate in log scale.

As a highly sensitive analysis tool, the uniformity and reproducibility of Raman signals are important influence factors for practical applications. We randomly selected 15 spots for statistics on the P–AgNPs@PMMA substrate. Figure 5a shows the SERS signals of the R6G molecules (10^{-7} M) collected on the P–AgNPs@PMMA substrate. It can be seen that the SERS spectra of R6G (10^{-7} M) from different points did not exhibit obvious changes in intensities or shapes. To more intuitively compare the peak fluctuation, the R6G Raman signal intensities at 614, 774, and 1362 cm^{-1} are shown in Figure 5b. These results demonstrate that the intensities of these peaks almost make a horizontal line, and the relative standard deviation of the peaks at 614, 774, and 1362 cm^{-1} are 6.913%, 3.926%, and 2.318%, respectively. These percentages are much lower than the scientific standard (20%), which indicates the high homogeneity and reproducibility of the P–AgNPs@PMMA substrate [31]. This is mainly because the AgNPs and PMMA composite structures provide uniform space distribution for analyte detection in designated hot spot zone, resulting in more intense electromagnetic field coupling, and a more uniform SERS signal. Besides, SERS active substrates not only need high sensitivity, high uniformity, but also good signal stability. Thus, time stability is another important parameter in the SERS detection.

We exposed the P–AgNPs@PMMA substrate to the air and performed a SERS test every five days to investigate the intensity change of the peak, as shown in Figure 5c. The Raman signal on the P–AgNPs@PMMA substrate drops very little. Figure 5d shows the corresponding Raman intensities of R6G molecule at 614 cm^{-1}. These experimental results demonstrate the P–AgNPs@PMMA hybrid detection geometry not only ensures uniform distribution of analyte molecules in the designated "hot spot" zone, but effectively prevents the embedded AgNPs from being oxidized by reacting with components in the air, as well as protect the target molecules from deformation and signal distortion caused by the direct interaction between AgNPs and molecules.

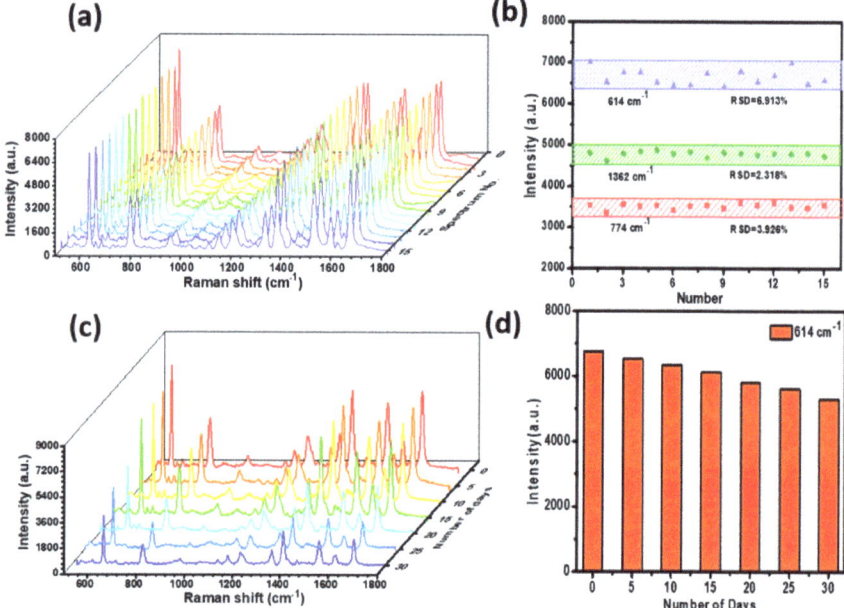

Figure 5. (**a**) Raman spectra collected from different points on the P–AgNPs@PMMA substrate for R6G molecules. The R6G concentration is measured at 10^{-7} M; (**b**) the peaks at 614, 774, and 1362 cm^{-1} demonstrate relative intensities collected from the Raman spectra; (**c**) the Raman spectra of the P–AgNPs@PMMA substrate was measured every five days at room temperature; (**d**) the Raman intensity of the 614 cm^{-1} peaks for R6G from (c).

Aside from the excellent homogeneity and stability, the P–AgNPs@PMMA substrate also exhibits good mechanical stability (stretching and bending) in practical applications. For tensile testing, the flexible substrate was stretched to ~110%, 120%, 130% (L/L_0; L: length after stretching; L_0: original length), as shown in Figure 6a. The flexible films were then bent in half with various bending cycles (Figure 6b). The optical image of stretching and bending are presented in Figure 6a$_1$, b$_1$. The Raman signals were analyzed for 10^{-7} M R6G molecules absorbed on the substrate after each mechanical stimulus (Figure.6a$_2$, b$_2$). In order to draw comparisons, the relative intensities of the characteristic peaks with error bars (same batch but different positions) at 614 and 774 cm^{-1} were collected to plot histograms for the stretching and bending processes, as displayed in Figure 6a$_3$, b$_3$. The histograms suggest that the Raman fingerprint peaks at 614 and 774 cm^{-1} remain mostly unchanged compared to the original state before mechanical stimulation. It can be seen that the SERS substrate can achieve mechanical deformation with little loss of SERS performance, and can better meet the requirements of SERS detection for uneven surfaces. This may be due to the good mechanical resistance of the PMMA film and the stability of the embedded AgNPs. It is well-known that the intensity of SERS is not only related to the distance of gaps, but also the number of molecules between gaps. During mechanical stimuli, the distance between silver nanoparticles becomes larger, which creates a large "hot spot" zone. The density of "hot spot" decreases within a certain incident laser area, while the number of adsorbed molecules increases in the "hot spot" zone. Therefore, mechanical changes in a certain range will not change the Raman intensity of the corresponding target molecule.

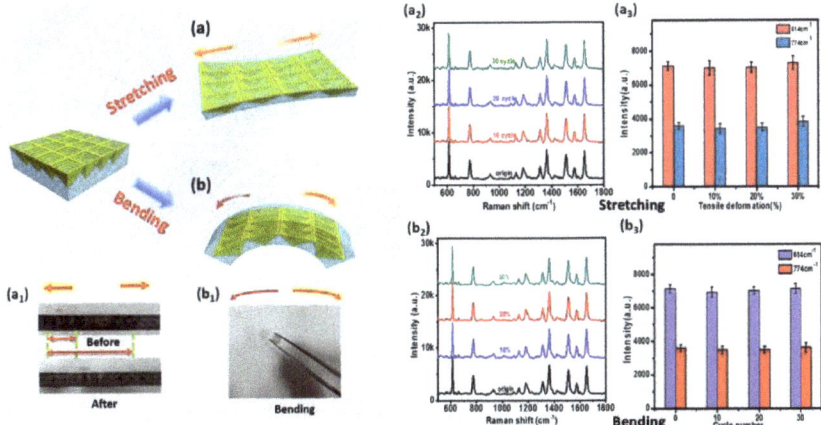

Figure 6. Durability tests with mechanical stimuli of the P-AgNPs@PMMA substrate. Schematic illustration: comparative Raman spectra and the SERS intensity of R6G at 614 and 774 cm^{-1} peaks; (**a**) after stretching the P-AgNPs@PMMA substrate to ~10%, 20%, 30%; (**b**) after the bending the P-AgNPs@PMMA substrate in half; (**a$_1$**) the optical image of stretching; (**b$_1$**) the optical image of bending; (**a$_2$**) raman signals of stretching; (**b$_2$**) raman signals of bending; (**a$_3$**) histograms for the stretching; (**b$_3$**) histograms for the bending.

During mechanical strain, stretching the P-AgNPs@PMMA substrate yields smaller gaps between neighboring particles, resulting in control over the electromagnetic fields coupling on a nanoscale [32]. To better understand the SERS behaviors of the flexible substrate during tensile testing, we employed commercial COMSOL software to model the local electric field properties of the flexible substrate. Herein, we model the electric field distributions of the SERS film to investigate the enhancement behavior. The height (~3 μm) and space (~4 μm) of the P-Si geometry was chosen according to the actual sample. The AgNPs embedded in the PMMA film is modeled as a sphere with a diameter of 78 nm and an interparticle distance of 15 nm, which mirrors the actual size in terms of SEM (Figure S1). Additionally, the incident light is 532 nm, and the refractive index of PMMA is 1.5, according to the actual experiments. Figure 7a–d respectively show the local electric field distributions at the y–z cross-section of the SERS film after stretching to ~0%, 10%, 20%, and 30%. The bottom left in Figure 7d exhibits the theoretical model of P-AgNPs@PMMA substrate. SEM image of the single pyramid-shaped AgNPs structure is also present in the inset of Figure 7d (bottom right). These results indicate that the electric field near the surface of the pyramid does not obviously change as the stretching length increases. Although some subtle differences have occurred, the plasmon coupling between high-density AgNPs embedded in the PMMA film remains unaffected, allowing for almost constant Raman signals, which is consistent with our experimental results.

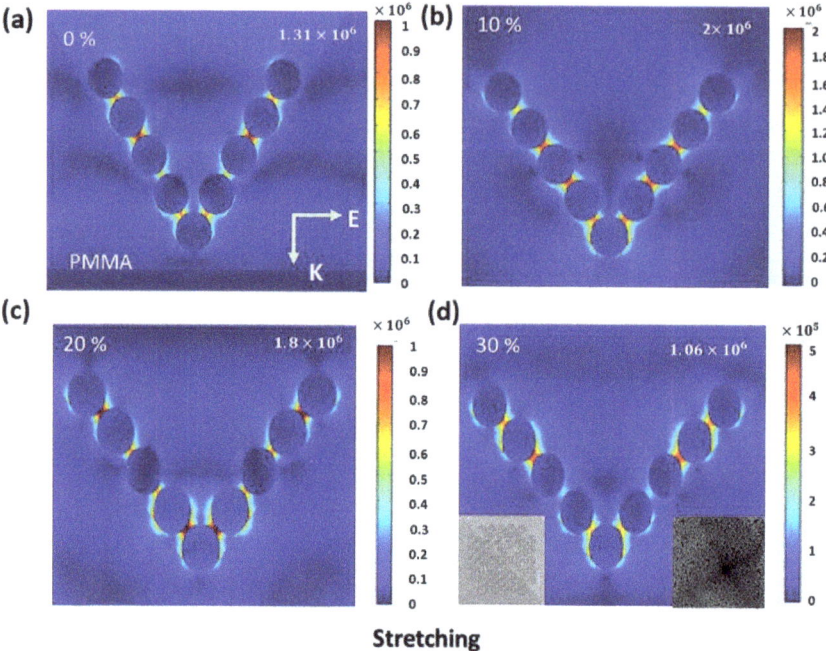

Figure 7. The respective Y–Z views of the electric filed distributed on the P–AgNPs@PMMA substrate after stretching to ~0% (**a**), 10% (**b**), 20% (**c**), and 30% (**d**).

To further investigate the practicability of the P–AgNPs@PMMA substrate, the in-situ detection of a solution containing biochemical molecules was performed. Adenosine is the metabolite of adenine nucleotides, which are regarded as major neuromodulators. Furthermore, as the core molecule of ATP and nucleic acids, adenosine forms a unique link among cell energy, gene regulation, and neuronal excitability [18,33]. Here, adenosine molecules were dissolved in deionized water, and the molecular concentration was measured at 10^{-8} M. The substrate was placed on the surface of the prepared adenosine aqueous solution. Next, in-situ detection was conducted, as exhibited in Figure 8a. The optical picture of the in-situ detection of adenosine molecules can be observed in Figure 8b. The detected results are presented in Figure 8c. There were no obvious Raman signals (orange line) obtained without the P–AgNPs@PMMA substrate. However, when the activated surface of the P–AgNPs@PMMA substrate encountered the solution, the characteristic peaks at 729, 1257, and 1329 cm^{-1} of adenosine molecules were easily detected (green line), which was attributed to the active surface of the substrate interacted with probe molecules, enhancing the Raman signal. The experimental results suggest the substrate's promising application to practical in-situ detection of biochemical molecules. Recently, MB is often used as the fish medicine or the disinfector for the fishpond. If MB molecules have not been completely removed from the skin of fish, it would be harmful to humans' health. We perform MB molecular detection by swabbing the surface of the fish skin using the P–AgNPs@PMMA substrate shown in Figure 8d. Figure 8e exhibits the SERS spectra of MB molecules. Obviously, the Raman intensity decrease with the decrease of the MB concentration. Based on these particle detection, the P–AgNPs@PMMA substrate has shown great potential in noninvasive and ultrasensitive molecular detention, such as the detecting on the surfaces with any arbitrary morphology and aqueous solution.

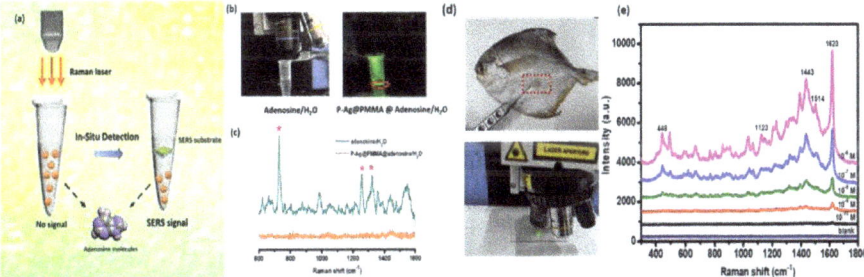

Figure 8. (**a**) Schematic illustration showing the in-situ detection process of the adenosine solution; (**b**) the optical picture of the in-situ detection of adenosine molecules; (**c**) Raman spectra of adenosine molecules before (orange line) and after (green line) placing the P–AgNPs@PMMA substrate on the adenosine solution to confirm the SERS effect. Insert: in-situ detection of adenosine molecule; (**d**) the optical picture of detecting of methylene-blue (MB) by swabbing the marine fish surface; (**e**) SERS spectra of MB obtained by swabbing fingerprint on the marine fish surface.

4. Conclusions

In summary, we proposed a rapid and convenient method for fabricating 3D P-AgNPs@PMMA flexible platforms, which enable highly sensitive single-molecule detection and provide a reproducible and stable Raman signal response. Benefiting from the effective oscillation of light inside the pyramid valley could provide wide distributions of 3D "hot spots" in a large space. The inclined surface design of the pyramid structure could facilitate the aggregation of probe molecules, achieving highly sensitive detection of R6G and CV. The AgNPs and PMMA composite structures provide uniform space distribution for analyte detection in designated hot spot zone, and the incident light can penetrate the external PMMA film to trigger the localized plasmon resonance of the encapsulated AgNPs, achieving enormous enhancement factor ($\sim 6.24 \times 10^8$). Additionally, the substrate maintains a stable SERS signal under various mechanical stimuli such as stretching and bending. As a practical application of the SERS substrate, we achieved the in-situ Raman detection of adenosine aqueous solution and the MB molecule detection of the skin of a fish. Our experimental results suggest promising application prospects for detection on the surfaces with any arbitrary morphology and aqueous solution.

Supplementary Materials: The following are available online at http://www.mdpi.com/2073-4360/12/2/392/s1, Figure S1: SEM image and the EDS spectrum of the P–AgNPs@PMMA substrate; Figure S2: SEM images of different volume ratio AgNPs@PMMA/P–Si substrate.

Author Contributions: Conceptualization, Y.J., Y.T. and H.W.; data curation, T.Z. and M.W.; writing—original draft preparation, M.W. and M.L.; writing—review and editing, C.L., Z.L. and M.L.; funding acquisition, C.Z., Q.S., B.M. and M.L.; All authors have read and agreed to the published version of the manuscript.

Funding: This research was funded by the National Natural Science Foundation of China (Grant No. 11774208 and 11974222) and Shandong Provincial Natural Science Foundation, China (Grant No. ZR2018MA040).

Conflicts of Interest: The authors declare no conflict of interest.

References

1. Han, X.X.; Zhao, B.; Ozaki, Y. Surface-enhanced Raman scattering for protein detection. *Anal. Bioanal. Chem.* **2009**, *394*, 1719–1727. [CrossRef] [PubMed]
2. Pieczonka, N.P.; Aroca, R.F. Single molecule analysis by surfaced-enhanced Raman scattering. *Chem. Soc. Rev.* **2008**, *37*, 946–954. [CrossRef] [PubMed]
3. Wang, Y.; Wei, H.; Li, B.; Ren, W.; Guo, S.; Dong, S.; Wang, E. SERS opens a new way in aptasensor for protein recognition with high sensitivity and selectivity. *Chem. Commun.* **2007**, *48*, 5220–5222. [CrossRef] [PubMed]
4. Wang, P.; Liang, O.; Zhang, W.; Schroeder, T.; Xie, Y.-H. Ultra-sensitive graphene-plasmonic hybrid platform for label-free detection. *Adv. Mater.* **2013**, *25*, 4918–4924. [CrossRef]

5. Liu, H.; Zhang, L.; Lang, X.; Yamaguchi, Y.; Iwasaki, H.; Inouye, Y.; Xue, Q.; Chen, M. Single molecule detection from a large-scale SERS-active Au 79 Ag 21 substrate. *Sci. Rep.* **2011**, *1*, 112. [CrossRef]
6. Lim, D.K.; Jeon, K.S.; Hwang, J.H.; Kim, H.; Kwon, S.; Suh, Y.D.; Nam, J.M. Highly uniform and reproducible surface-enhanced Raman scattering from DNA-tailorable nanoparticles with 1-nm interior gap. *Nat. Nanotechnol.* **2011**, *6*, 452–460. [CrossRef]
7. Li, J.F.; Huang, Y.F.; Ding, Y.; Yang, Z.L.; Li, S.B.; Zhou, X.S.; Fan, F.R.; Zhang, W.; Zhou, Z.Y.; Wu, D.Y.; et al. Shell-isolated nanoparticle-enhanced Raman spectroscopy. *Nature* **2010**, *464*, 392–395. [CrossRef]
8. Fang, Y.; Seong, N.H.; Dlott, D.D. Measurement of the distribution of site enhancements in surface-enhanced Raman scattering. *Science* **2008**, *321*, 388–392. [CrossRef]
9. Lal, S.; Grady, N.K.; Kundu, J.; Levin, C.S.; Lassiter, J.B.; Halas, N.J. Tailoring plasmonic substrates for surface enhanced spectroscopies. *Chem. Soc. Rev.* **2008**, *37*, 898–911. [CrossRef]
10. Pedano, M.L.; Li, S.; Schatz, G.C.; Mirkin, C.A. Periodic electric field enhancement along gold rods with nanogaps. *Angew. Chem. Inter. Ed.* **2010**, *49*, 78–82. [CrossRef]
11. Wu, L.; Wang, Z.; Shen, B. Large-scale gold nanoparticle superlattice and its SERS properties for the quantitative detection of toxic carbaryl. *Nanoscale* **2013**, *5*, 5274–5278. [CrossRef] [PubMed]
12. Jiang, J.; Zou, S.; Ma, L.; Wang, S.; Liao, J.; Zhang, Z. Surface-enhanced Raman scattering detection of pesticide residues using transparent adhesive tapes and coated silver nanorods. *ACS Appl. Mater. Inter.* **2018**, *10*, 9129–9135. [CrossRef] [PubMed]
13. Park, S.; Lee, J.; Ko, H. Transparent and flexible surface-enhanced Raman scattering (SERS) sensors based on gold nanostar arrays embedded in silicon rubber film. *ACS Appl. Mater. Inter.* **2017**, *9*, 44088–44095. [CrossRef] [PubMed]
14. Zhao, Y.; Yang, D.; Li, X.; Liu, Y.; Hu, X.; Zhou, D.; Lu, Y. Toward highly sensitive surface-enhanced Raman scattering: The design of a 3D hybrid system with monolayer graphene sandwiched between silver nanohole arrays and gold nanoparticles. *Nanoscale* **2017**, *9*, 1087–1096. [CrossRef] [PubMed]
15. Yang, C.; Chen, Y.; Liu, D.; Chen, C.; Wang, J.; Fan, Y.; Huang, S.; Lei, W. Nanocavity-in-multiple nanogap plasmonic coupling effects from vertical sandwich-like Au@ Al_2O_3@ Au arrays for surface-enhanced Raman scattering. *ACS Appl. Mater. Inter.* **2018**, *10*, 8317–8323. [CrossRef] [PubMed]
16. Kumar, P.; Khosla, R.; Soni, M.; Deva, D.; Sharma, S.K. A highly sensitive, flexible SERS sensor for malachite green detection based on Ag decorated microstructured PDMS substrate fabricated from Taro leaf as template. *Sens. Actuators B-Chem.* **2017**, *246*, 477–486. [CrossRef]
17. Kumar, S.; Lodhi, D.K.; Goel, P.; Mishra, P.; Singh, J. A facile method for fabrication of buckled PDMS silver nanorod arrays as active 3D SERS cages for bacterial sensing. *Chem. Commun.* **2015**, *51*, 12411–12414. [CrossRef]
18. Qiu, H.; Wang, M.; Jiang, S.; Zhang, L.; Yang, Z.; Li, L.; Li, J.; Cao, M.; Huang, J. Reliable molecular trace-detection based on flexible SERS substrate of graphene/Ag-nanoflowers/PMMA. *Sens. Actuators B-Chem.* **2017**, *249*, 439–450. [CrossRef]
19. Zhang, C.; Jiang, S.; Huo, Y.; Liu, A.; Xu, S.; Liu, X.; Sun, Z.; Xu, Y.; Li, Z.; Man, B. SERS detection of R6G based on a novel graphene oxide/silver nanoparticles/silicon pyramid arrays structure. *Opt. Express* **2015**, *23*, 24811–24821. [CrossRef]
20. Zhang, F.; Braun, G.B.; Shi, Y.; Zhang, Y.; Sun, X.; Reich, N.O.; Zhao, D.; Stucky, G. Fabrication of Ag@ SiO_2@Y_2O_3: Er nanostructures for bioimaging: Tuning of the upconversion fluorescence with silver nanoparticles. *J. Am. Chem. Soc.* **2010**, *132*, 2850–2851. [CrossRef]
21. Zhao, J.; Jensen, L.; Sung, J.; Zou, S.; Schatz, G.C.; Van Duyne, R.P. Interaction of plasmon and molecular resonances for rhodamine 6G adsorbed on silver nanoparticles. *J. Am. Chem. Soc.* **2007**, *129*, 7647–7656. [CrossRef] [PubMed]
22. Michaels, A.M.; Jiang, J.; Brus, L. Ag nanocrystal junctions as the site for surface-enhanced Raman scattering of single rhodamine 6G molecules. *J. Phys. Chem. B* **2000**, *104*, 11965–11971. [CrossRef]
23. Liu, M.; Shi, Y.; Zhang, G.; Zhang, Y.; Wu, M.; Ren, J.; Man, B. Surface-enhanced Raman spectroscopy of two-dimensional tin diselenide nanoplates. *Appl. Spectrosc.* **2018**, *72*, 1613–1620. [CrossRef] [PubMed]
24. Hakonen, A.; Svedendahl, M.; Ogier, R.; Yang, Z.J.; Lodewijks, K.; Verre, R.; Shegai, T.; Andersson, P.O.; Kall, M. Dimer-on-mirror SERS substrates with attogram sensitivity fabricated by colloidal lithography. *Nanoscale* **2015**, *7*, 9405–9410. [CrossRef] [PubMed]

25. Qian, Y.; Meng, G.; Huang, Q.; Zhu, C.; Huang, Z.; Sun, K.; Chen, B. Flexible membranes of Ag-nanosheet-grafted polyamide-nanofibers as effective 3D SERS substrates. *Nanoscale* **2014**, *6*, 4781–4788. [CrossRef]
26. Lee, C.H.; Tian, L.; Singamaneni, S. Based SERS swab for rapid trace detection on real-world surfaces. *ACS Appl. Mater. Inter.* **2010**, *2*, 3429–3435. [CrossRef] [PubMed]
27. Zhou, Y.; Cheng, X.; Yang, J.; Zhao, N.; Ma, S.; Li, D.; Zhong, T. Fast and green synthesis of flexible free-standing silver nanoparticles–graphene substrates and their surface-enhanced Raman scattering activity. *RSC Adv.* **2013**, *3*, 23236–23241. [CrossRef]
28. Zhang, W.; Li, B.; Chen, L.; Wang, Y.; Gao, D.; Ma, X.; Wu, A. Brushing, a simple way to fabricate SERS active paper substrates. *Analy. Methods* **2014**, *6*, 2066–2071. [CrossRef]
29. Fortuni, B.; Fujita, Y.; Ricci, M.; Inose, T.; Aubert, R.; Lu, G.; Hutchison, J.A.; Hofkens, J.; Latterini, L.; Uji, I.H. A novel method for in situ synthesis of SERS-active gold nanostars on polydimethylsiloxane film. *Chem. Commun. (Camb.)* **2017**, *53*, 5121–5124. [CrossRef]
30. Loera-Serna, S.; Ortiz, E.; Beltrán, H.I. First trial and physicochemical studies on the loading of basic fuchsin, crystal violet and Black Eriochrome T on HKUST-1. *New J. Chem.* **2017**, *41*, 3097–3105. [CrossRef]
31. Natan, M.J. Concluding remarks surface enhanced Raman scattering. *Faraday discuss.* **2006**, *132*, 321–328. [CrossRef] [PubMed]
32. Cataldi, U.; Caputo, R.; Kurylyak, Y.; Klein, G.; Chekini, M.; Umeton, C.; Bürgi, T. Growing gold nanoparticles on a flexible substrate to enable simple mechanical control of their plasmonic coupling. *J. Mater. Chem. C* **2014**, *2*, 7927–7933. [CrossRef]
33. Xu, S.; Man, B.; Jiang, S.; Wang, J.; Wei, J.; Xu, S.; Liu, H.; Gao, S.; Liu, H.; Li, Z.; et al. Graphene/Cu nanoparticle hybrids fabricated by chemical vapor deposition as surface-enhanced Raman scattering substrate for label-free detection of adenosine. *ACS appl. Mater. Inter.* **2015**, *7*, 10977–10987. [CrossRef] [PubMed]

© 2020 by the authors. Licensee MDPI, Basel, Switzerland. This article is an open access article distributed under the terms and conditions of the Creative Commons Attribution (CC BY) license (http://creativecommons.org/licenses/by/4.0/).

Article

Magnetite Nanoparticles Coated with PEG 3350-Tween 80: In Vitro Characterization Using Primary Cell Cultures

Jorge A Roacho-Pérez [1], Fernando G Ruiz-Hernandez [1], Christian Chapa-Gonzalez [2], Herminia G Martínez-Rodríguez [1], Israel A Flores-Urquizo [3], Florencia E Pedroza-Montoya [1], Elsa N Garza-Treviño [1], Minerva Bautista-Villareal [4], Perla E García-Casillas [2,*] and Celia N Sánchez-Domínguez [1,*]

[1] Departamento de Bioquímica y Medicina Molecular, Facultad de Medicina, Universidad Autónoma de Nuevo León, Monterrey, Nuevo León 64460, Mexico; jorge.roacho@uacj.mx (J.A.R.-P.); fernandogruiz@outlook.com (F.G.R.-H.); herminiamar@gmail.com (H.G.M.-R.); florencia.estefana@gmail.com (F.E.P.-M.); egarza.nancy@gmail.com (E.N.G.-T.)
[2] Instituto de Ingeniería y Tecnología, Universidad Autónoma de Ciudad Juárez, Ciudad Juárez, Chihuahua 32310, Mexico; christian.chapa@uacj.mx
[3] Facultad de Ciencias Químicas, Universidad Autónoma de Nuevo León, San Nicolas de los Garza, Nuevo León 66455, Mexico; Israel.flores.ur@gmail.com
[4] Departamento de Ciencias de los Alimentos, Facultad de Ciencias Biológicas, Universidad Autónoma de Nuevo León, San Nicolas de los Garza, Nuevo León 66455, Mexico; lca.minevillarreal90@gmail.com
* Correspondence: pegarcia@uacj.mx (P.E.G.-C.); celia.sanchezdm@uanl.edu.mx (C.N.S.-D.)

Received: 30 December 2019; Accepted: 24 January 2020; Published: 2 February 2020

Abstract: Some medical applications of magnetic nanoparticles require direct contact with healthy tissues and blood. If nanoparticles are not designed properly, they can cause several problems, such as cytotoxicity or hemolysis. A strategy for improvement the biological proprieties of magnetic nanoparticles is their functionalization with biocompatible polymers and nonionic surfactants. In this study we compared bare magnetite nanoparticles against magnetite nanoparticles coated with a combination of polyethylene glycol 3350 (PEG 3350) and polysorbate 80 (Tween 80). Physical characteristics of nanoparticles were evaluated. A primary culture of sheep adipose mesenchymal stem cells was developed to measure nanoparticle cytotoxicity. A sample of erythrocytes from a healthy donor was used for the hemolysis assay. Results showed the successful obtention of magnetic nanoparticles coated with PEG 3350-Tween 80, with a spherical shape, average size of 119.2 nm and a zeta potential of +5.61 mV. Interaction with mesenchymal stem cells showed a non-cytotoxic propriety at doses lower than 1000 µg/mL. Interaction with erythrocytes showed a non-hemolytic propriety at doses lower than 100 µg/mL. In vitro information obtained from this work concludes that the use of magnetite nanoparticles coated with PEG 3350-Tween 80 is safe for a biological system at low doses.

Keywords: nanoparticles; polyethylene glycol; Tween 80; cytotoxicity; hemotoxicity; primary cell culture; medical applications

1. Introduction

Interest in nanotechnology has increased in recent years, looking for its application in different areas, such as agriculture, environmental sciences or even as a solution to various emerging diseases in medicine. Currently, magnetic nanoparticles (MNPs) are one of the most studied nanoparticles in the biomedical field [1,2]. These kinds of nanoparticles are easily manipulated in size, shape and chemical properties. They also have unique physical properties, are biocompatible with the human body, and have a low production cost [3]. MNPs must be designed taking into consideration their physical characteristics to be biocompatible with blood elements and not cause cytotoxicity to other tissues. Nanoparticle size should be from 30 to 150 nm because nanoparticles with smaller sizes are immediately discarded from the blood by the kidneys and large sizes hardly cross biological barriers like epithelia [4]. Due to these characteristics, MNPs have a potential to be applied as an alternative in the diagnosis or treatment of several diseases, such as cancer.

Although MNPs have given good results in preclinical and clinical trials, they can be coated with other materials to improve their proprieties [5–8]. A coat of polymeric materials is preferred because of their proprieties of biodegradation and biocompatibility [9]. However, a common problem with some polymeric nanoparticles in blood is their high surface charge, which can interact with erythrocytes causing lysis. Polyethylene glycol (PEG) is a nontoxic and non-electrostatic-charge linear polymer [4] that has already been used to coat MNPs. Although there is a common use of PEG in medicine [10,11], the interaction of the PEG molecule with the human body is still being studied [12–14]. According to several authors, coating MNPs with PEG improves the MNP blood circulation and their biocompatibility; it reduces their surface charge and maintains the magnetic characteristics of the MNPs [1,5,9]. On the other hand, Tween 80 has been proposed by some authors such as a coat in nanoparticles because of its ability to be targeted into the brain after an intravenous injection. Tween 80 is a biocompatible non-ionic surfactant that binds with the plasma lipoprotein Apo-E, which attaches to Low-Density Lipoprotein (LDL) receptors on the brain micro vascular endothelial cells. This propriety can be used by the nanoparticles in order to delivery substances such as drugs into the brain [15,16]. Tween 80 has also been proposed for the delivery of drugs into cancerous cells [17], has been used as a coating for metallic nanoparticles [18,19], and is widely used in cosmetics, food and pharmaceutics. However, the safety of PEG and Tween 80, avoid the possible harm of nanoparticles is still a challenge for nanomedicine developers. It is a priority to analyze the cytotoxicity of nanoparticles with a focus on the identification of a safe dose range.

Different methodologies for the synthesis of nanoparticles generate nanoparticles with different physical proprieties (size, Z potential, morphology) that leads distinct biological behaviors (cytotoxicity and hemotoxicity). So, for researchers who work in the development of nanoparticles for biomedical applications is important to evaluate the in vitro biological safety of particles to ensure there will be no damage at cellular level. In this study, an in vitro biological characterization was performed with magnetite nanoparticles coated with PEG 3350-Tween 80. Nanoparticle synthesis was developed in order to obtain nanoparticles with physical proprieties, such as size, shape and z-potential, needed for biomedical application. For the evaluation of the nanoparticle interaction with cells, a mesenchymal stem cell (MSC) primary culture from sheep adipose tissue was exposed to different concentrations of magnetic nanoparticles. For cytotoxic assays, primary cell cultures are more sensitive, this is why they are better test subjects than a transformed cell line. The use of undifferentiated cells provides a broader picture of the cytotoxic effects of the nanoparticles on different tissues. Human blood cells from a healthy donor were exposed to the nanoparticles to evaluate hemotoxicity in erythrocytes (hemolysis test) based on the measurement of the lysis of erythrocytes.

2. Materials and Methods

This project was reviewed and approved with the identification code BI17-00001 by the ethics in research committee of the School of Medicine and Dr. José Eleuterio Gonzalez University Hospital of the Universidad Autónoma de Nuevo León, approved in September 2017. All animals and human donors were treated according to ethical standards.

2.1. Magnetite Nanoparticle Synthesis

Magnetite nanoparticles were synthetized by chemical co-precipitation based on the modified methodology of Chapa-González et al. [20]. We used two precursor solutions: 100 mL of ferric chloride hexahydrate 0.01 M ($FeCl_3 \cdot 6H_2O$, Thermo Fisher Scientific, Waltham, MA, USA) and 100 mL of ferrous sulfate heptahydrated 0.05 M ($FeSO_4 \cdot 7H_2O$, Thermo Fisher Scientific, Waltham, MA, USA).

Solutions were mixed, agitated and heated until reaching 50 °C. An aliquot of 100 mL of a strong base solution of ammonium hydroxide (NH$_4$OH, CTR Scientific, Monterrey, Mexico) was added abruptly to increase the medium pH to produce precipitation of the nanoparticles. The solution was then agitated and heated until it reached 80 °C. The precipitate was washed with distillated water and was isolated by centrifugation three times at 5000 rpm for 10 min until nanoparticles reached a pH of 7. Magnetite nanoparticles were dried at 50 °C for 24 h, and then they were pulverized and saved for functionalization.

2.2. Functionalization with PEG 3350-Tween 80

Figure 1 shows the coating process of magnetite nanoparticles. A 25 mg sample of magnetite nanoparticles was dispersed in 25 mL of a 7% Tween 80 (Sigma-Aldrich, St. Louis, MO, USA) solution and 50 mg of PEG 3350 (Thermo Fisher Scientific, Waltham, MA, USA), mixing it with an ultrasonic processor (Fisher Scientific, Pittsburgh, PA, USA); amplitude 80%. The pH was adjusted with ammonium hydroxide until it reached a pH of 9. Nanoparticles were isolated by centrifugation at 3000 rpm for 5 min and washed with distilled water until nanoparticles reached a pH of 7. Nanoparticles were dried at 50 °C for 24 h.

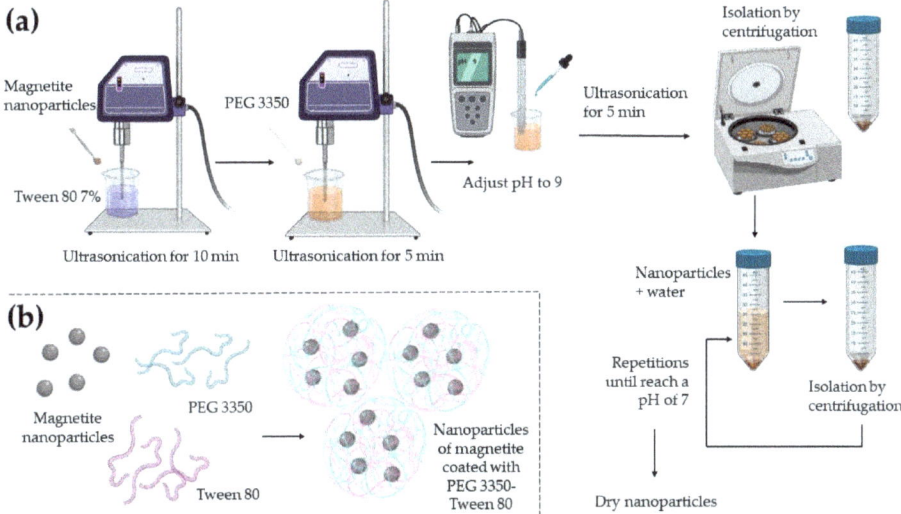

Figure 1. A scheme cartoon of the coating process of nanoparticles with PEG 3350 and Tween 80. (a) A complete diagram of the coating process methodology. (b) The interaction of PEG 3350 and Tween 80 with the magnetite nanoparticles is possible because of physical forces, creating a polymer network around the magnetite nanoparticles.

2.3. Nanoparticle Characterization

An X-ray diffraction analysis (XDR D2 PHASER; Bruker, Billerica, MA, USA) was performed for the nanoparticles of magnetite. Sample's XRD pattern was recorded with a LYNXEYE XE-T detector over the 2θ range from 10 to 90° at a scan rate of 0.05°/0.5 s employing Cu-Kα radiation (1.5408 Å). The formed phase was analyzed using a JCPDS PDF card number 01-088-0315 corresponding to magnetite (Fe$_3$O$_4$). MNPs coat was evaluated by Fourier transform infrared spectroscopy (FTIR NICOLET 6700, Thermo Fisher Scientific, Waltham, MA, USA), to analyze functional groups present in the samples. The measures were carried out in the range from 600 to 4000 cm^{-1} with a 2 cm^{-1} resolution for 130 scanning times.

Nanoparticle morphology and size were examined with scanning electron microscopy (FE-SEM SU5000; HITACHI, Tokyo, Japan). A sample of 3 mg of nanoparticles was dispersed in 3 mL of ultrapure water using an ultrasonic processor. A drop of nanoparticles solution was spread on a carbon tape slide. The sample was dried and subsequently a SEM and an energy-dispersive X-ray spectroscopy (EDS) study were performed. The zeta potential and size distribution in water was measured with dynamic light scattering (DLS) equipment (Zetasizer Nano ZS90, Malvern Instruments, Malvern, UK). The measurements were carried out using a universal dip cell (ZEN 1002, Malvern Instrument, Malvern, UK) at 25 °C. According to literature, nanoparticle samples were dispersed (1000 µg/mL) into a previously filtered ultrapure water immediately before their analysis [21].

2.4. Obtention of an MSC Primary Culture

An MSC primary cell culture from sheep adipose tissue was obtained and characterized. The tissue sample was provided by the School of Veterinary Medicine and Zootechnics of the Universidad Autonoma de Nuevo Leon. It came from an independent chirurgical procedure from this study. Subcutaneous fat tissue was obtained from the abdominal region of the sheep. The tissue was transported with supplemented sterile PBS inside a cooler with ice until its processing in a safety cabinet in the laboratory. The supplemented PBS contained PBS pH 7.4 (Thermo Fisher Scientific, Waltham, MA, USA), 0.5% of Gentamicin at 10 mg/mL (Thermo Fisher Scientific, Waltham, MA, USA), 0.5% of Penicillin-Streptomycin at 10 mg/mL (Thermo Fisher Scientific, Waltham, MA, USA) and 0.1% of Fungizone at 250 µg/mL (Thermo Fisher Scientific, Waltham, MA, USA). Fat tissue was washed three times with PBS and three more times with PBS supplemented. With the help of sterile surgical tools, the tissue was fragmented to obtain small portions (mechanical disintegration). Next, a volume of the fragmented tissue was enzymatically disaggregated in two volumes of a 0.2% collagenase I solution. Collagenase I solution contained PBS, 1.2% of Collagenase Type I at 285 U/mg (Thermo Fisher Scientific, Waltham, MA, USA), 0.5% of Gentamicin at 10 mg/mL and 0.1% of Fungizone at 250 µg/mL. The incubation in the collagenase I solution was performed in agitation at 37 °C from 2 to 2.5 h until observing the disintegration of the tissue. The traces of non-disaggregated tissue were eliminated, and individual cells were centrifuged at 5000 rpm for 10 min. The cell button was washed with PBS and finally, resuspended in 1 mL of supplemented MEM medium. The supplemented MEM medium contained Minimum essential medium alpha +GlutaMAX (Thermo Fisher Scientific, Waltham, MA, USA), 10% of SBF (Thermo Fisher Scientific, Waltham, MA, USA), 0.5% of Gentamicin at 10 mg/mL and 0.1% of Fungizone at 250 µg/mL. Cells were cultured for 3-5 days at 37 °C and 5% of CO_2 until obtained adherent cells with fibroblastic morphology.

In order to be sure that the cells extracted were MSCs, a characterization by immunocytochemistry was performed. CD-90 goat monoclonal c2441-60 (USBiological, San Antonio, TX, USA) and CD105 mouse monoclonal c2446-55 (USBiological, San Antonio, TX, USA) antibodies were used for immunochemical characterization, using a mouse and rabbit specific HRP/DAB detection system ab64264 (Abcam, Cambridge, UK).

2.5. Cytotoxicity: MTT Assay

To assess whether MNPs could cause damage that affects cell viability, an MTT assay were realized. A total of 5000 MSC were seeded per well in a 96 well microplate and cultured with complemented DMEM. Complemented DMEM contained Dulbecco's Modified Eagle's Medium (Thermo Fisher Scientific, Waltham, MA, USA), 10% of SBF and 1% of Penicillin-Streptomycin. After 24 h, cells were exposed with the nanoparticles previously sterilized by UV light exposition. The concentrations evaluated by triplicated were a logarithmic series from 10 to 10,000 µg/mL. Cell death was measured after 24 h of exposition. For MTT assay, cells were incubated with an MTT solution for 4 h. The MTT solution contained MTT (Sigma-Aldrich, St. Louis, MO, USA) at 100 µg/mL in complemented DMEM. After incubation, medium was retired, and formazan was dissolved in 50 µL of isopropanol pH 3. Absorbance was read at 570 nm.

2.6. Hemolysis Test

To measure the impact of nanoparticles in erythrocyte lysis, a quintupled hemolysis test was performed according to Macías-Martínez et al. methodology [22]. Nanoparticles at different concentrations (1, 10, 100 and 1000 µg/mL) were dispersed in Alsever's solution (dextrose 0.116 M, sodium chloride 0.071 M, sodium citrate 0.027 M and citric acid 0.002 M, pH 6.4). Moreover, a total of 5 mL of blood from a healthy donor was collected in a heparinized-tube (Becton Dickinson, Franklin Lakes, NJ, USA). Erythrocytes were separated by centrifugation (3000 rpm for 4 min) and washed three times using an Alsever's solution. Nanoparticles and erythrocytes were incubated in a relation of 1:99 erythrocytes: nanoparticles v/v, for 30 min, at 37 °C in agitation at 400 rpm. For positive and negative controls, a suspension of erythrocytes with distilled water and Alsever's solution, respectively, was used (relation 1:99 v/v). After the incubation, samples were centrifuged at 3000 rpm for 4 min. Hemoglobin released in the supernatant was measured by UV-Vis spectroscopy (NanoDrop, Thermo Fisher Scientific, Waltham, MA, USA) at 415 nm. A positive control with distilled water was considered as 100% hemolysis. Mathematical calculation was required to determine the percentage of hemolysis of problem samples.

2.7. Statistical Analysis

For MTT and hemolysis assays a statistical analysis was developed. The cell viability percentages, as well as percentages of hemolysis, were analyzed with an analysis of variance (ANOVA) and a Tukey's HSD (Honestly Significant Difference) test, with a confidence interval of 95% to determine significant differences between the control group and the test samples.

3. Results and Discussion

3.1. Magnetite Nanoparticle Synthesis

Magnetite nanoparticles are nanoferrites characterized by having a crystalline structure. Magnetite (Fe_3O_4) is an iron mineral that is composed of a combination of oxide Fe^{2+} and Fe^{3+}. Magnetite nanoparticle synthesis was performed by the nanoprecipitation method, where the chemical reaction that takes place for the formation of magnetite is represented by the Equation (1) [20]:

$$Fe^{2+} + 2Fe^{3+} + 8OH^- \rightarrow Fe_3O_4 + 4H_2O \quad (1)$$

The XRD pattern was employed to determine crystal structure and size of the synthesized magnetite nanoparticles. Figure 2a shows the magnetite nanoparticle XRD pattern. It confirms the formation of magnetite in a pure phase because there are no other peaks aside from the magnetite phase. Additionally, according to consulted literature those diffraction peaks correspond to the cubic spinel structure of magnetite (Fe_3O_4). Authors report the same diffraction peaks for the magnetite elaborated via co-precipitation [23–28]. The determination of the average crystallite size was estimated by the Debye–Scherrer Equation (2)

$$D = 0.9\lambda/\beta\cos(\theta) \quad (2)$$

where λ is the wavelength (1.5408 Å), β is the full width at half maximum of the most intense peak (3 1 1) and θ is the diffraction angle of the (3 1 1) peak (35.61 degrees). The sample's average crystallite size calculated was 9.49 nm. The obtained size is comparable with the references [23,28]. The FTIR spectra of magnetite was recorded and displayed in the Figure 2b, this analysis was done to verify the presence of characteristics bands of magnetite [29]. Bare magnetite nanoparticles show one main band also reported by other authors, corresponding to the presence of Fe–O [23,27,30]. Both, the XRD and the FTIR analyzes, provide information about a reliable methodology for the synthesis of magnetite nanoparticles.

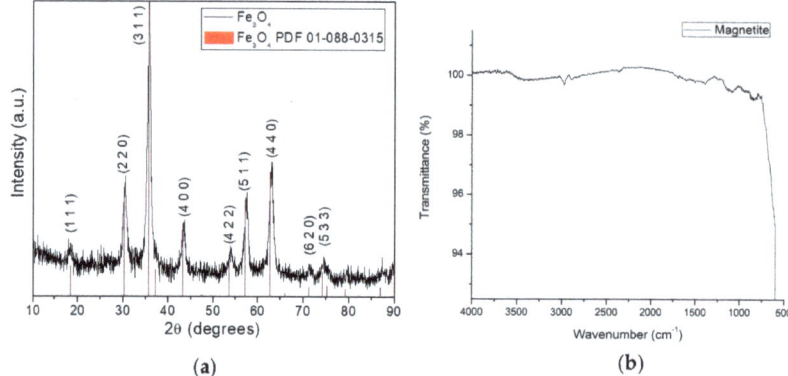

Figure 2. Magnetic nanoparticles (MNPs) synthesis characterization. (**a**) XRD pattern from magnetite nanoparticles exhibits nine diffraction peaks observed at 2θ value 18.29, 30.13, 35.61, 43.26, 53.84, 57.32, 63.13, 71.15 and 74.17, which match with the reference PDF 01-088-0315 card. Those peaks are indexed to the (1 1 1), (2 2 0), (3 1 1), (4 0 0), (4 2 2), (5 1 1), (4 4 0), (6 2 0) and (5 3 1) planes respectively; (**b**) FTIR spectra of bare magnetite. Magnetite shows the first band around 600 cm^{-1}. It is related to Fe–O stretching in the octahedral site of the crystal structure of magnetite.

3.2. Magnetite Coating with PEG 3350-Tween 80

PEGylation consists in a polymeric conjugation strategy. Due to PEG have two hydroxyl groups at their ends, PEG can adhere to almost any structure [10]. The FTIR spectra of magnetite coated with PEG 3350-Tween 80 was recorded and displayed in the Figure 3. This analysis was done to verify the presence of characteristics bands of magnetite on the sample and PEG 3350-Tween 80 on its surface. The spectrum presents more bands than the bare magnetite spectrum due to different functional groups present in PEG 3350 and in Tween 80. The spectrum shown the vibrational modes corresponding to the vibration modes of magnetite, PEG 3350 and Tween 80. So, was confirmed that magnetite nanoparticles were coated with PEG successfully. The identified vibration modes of PEG and Tween 80 are also described in other reports [27,31–36].

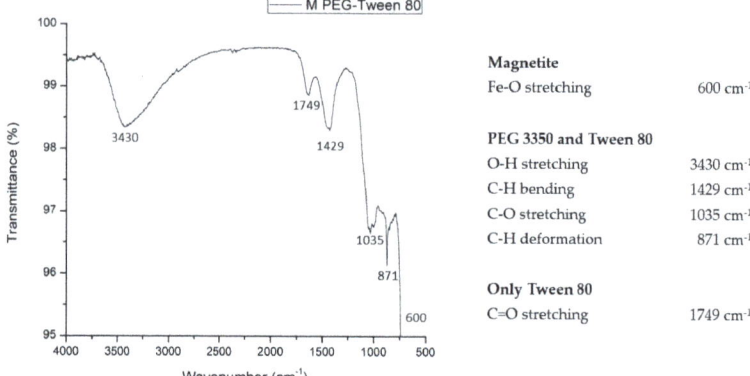

Figure 3. FTIR spectra of magnetite coated with PEG 3350-Tween 80. Obtained bands correspond to bonds found in PEG 3350 as well as in Tween 80: deformation (871 cm^{-1}) and bending (1429 cm^{-1}) vibration of C–H; C–O stretching (1035 cm^{-1}) and the O-H stretching (3430 cm^{-1}). Samples present the Fe–O stretching (600 cm^{-1}) coming from the magnetite and the C=O stretching (1749 cm^{-1}) coming from the Tween 80.

3.3. Nanoparticle Characterization: Shape

Properties of nanoparticles changes when they are functionalized, so they need to be evaluated [37]. Physical characteristics of the nanoparticles are important to estimate the efficiency of a nanoparticle. Morphology is a propriety that affect in the efficiency of the nanoparticles internalization into cells is nanoparticle geometric morphology [38]. Shape is also correlated with lifetime in blood circulation [39]. Magnetite (Figure 4a) and magnetite nanoparticles coated with PEG 3350-Tween 80 (Figure 4b–d) SEM images are shown. Coated nanoparticles shown a spherical shape. In addition, EDS analysis showed the chemical elements that compound the nanoparticles coated with PEG 3350-Tween80, demonstrating the presence of Fe coming from magnetite inside the samples.

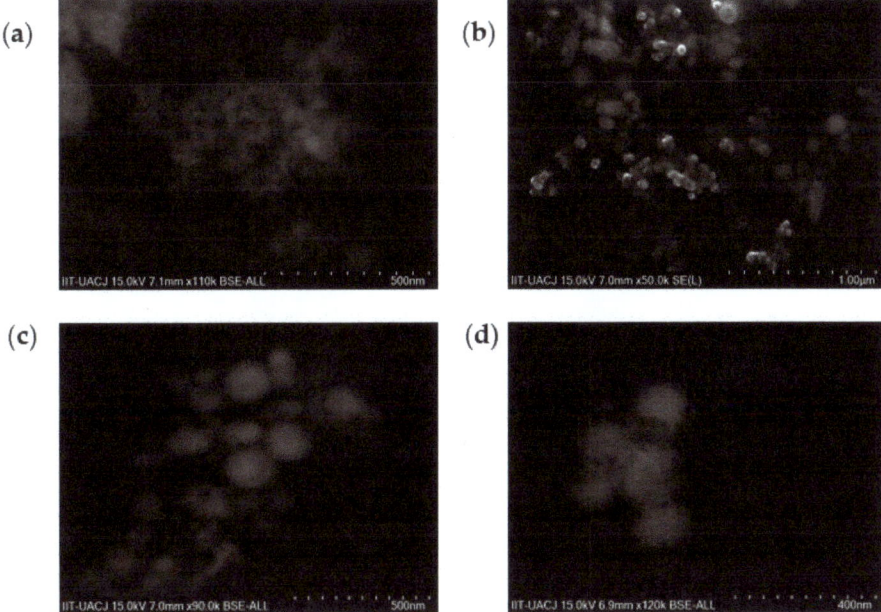

Figure 4. SEM images. (**a**) Magnetite nanoparticles and (**b–d**) magnetite nanoparticles coated with PEG 3350-Tween 80 at different magnifications. All the samples showed spherical shapes. In bare magnetite nanoparticles are evidence of agglomeration, while in magnetite nanoparticles coated with PEG 3350-Tween 80 the separation between one nanoparticle and other becomes more evident.

3.4. Nanoparticle Characterization: Size

DLS technique was used to analyze the size of the nanoparticles. This technique measures the size of nanoparticles dispersed in an aqueous solution, so if the nanoparticles generates agglomerates, the given size is correlated with the size of the agglomerates [9]. DLS of bare magnetite (Figure 5a) shows single particle average size of 10.51 nm and agglomerates of 527.8 and 4367 nm. The sizes of the bare magnetite single nanoparticle correspond to their average crystallite size given by XRD, so this evidence indicates that magnetite nanoparticles generated are monocrystalline. Although, only the 2.6% of the sample generate agglomerates of almost 4.4 µm, biggest agglomerates can cause several problems if they are administrated by blood into the patient. DLS results of magnetite nanoparticles coated with PEG 3350-Tween 80 (Figure 5b) shows single particle average size of 119.2 nm and agglomerates average size of 785.6 nm. Although, the agglomerates generated by coated nanoparticles are in a bigger proportion than bare nanoparticles, the sizes of the biggest agglomerates of coated nanoparticles (2.7 µm) are approximately 2.3 times smaller than the size of bare nanoparticle agglomerates (6.4 µm). The size of the nanoparticles and their agglomerates affects their elimination and circulation time in the organism. Sizes from 30 to 100 nm could penetrate well into highly permeable tumors [40]. Nanoparticles with a diameter smaller than 5.5 nm results in a rapid urinary excretion [41]. A disadvantage of larger particles is that they are easily recognized by the macrophages from the immune system, while smaller sizes have a longer circulation time in blood [4]. In fact, agglomeration of coated nanoparticles is demonstrated in DLS results, but SEM images shows well defined nanoparticles in the nanometer scale. These agglomerates can be avoided by the application of a mechanical force such as ultrasonication. Magnetite nanoparticles coated with PEG 3350-Tween 80 single nanoparticle have the optimum size because they are large enough for not to be eliminated from the bloodstream due to renal excretion, but, are small enough to cross epithelia and not be quickly eliminated by the macrophages.

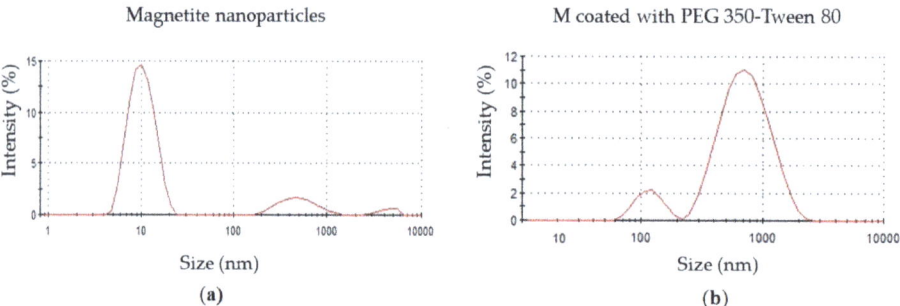

Figure 5. Dynamic light scattering (DLS) sizes results. (**a**) Bare magnetite nanoparticles size distribution. With an average size of 10.51 nm 83.8% of the nanoparticles do not generate agglomerates. The 13.6% of the nanoparticles generates agglomerates of 527.8 nm and the other 2.6% generates agglomerates of 4367 nm. (**b**) Magnetite nanoparticles coated with PEG 3350-Tween 80 size distribution. With an average size of 119.2 nm 10.3% of the nanoparticles do not generate agglomerates. The 89.7% of the nanoparticles generates agglomerates of 785.6 nm.

3.5. Nanoparticle Characterization: Zeta Potential

DLS showed a zeta potential of −3.41 mV for magnetite nanoparticles and a value of +5.61 mV for magnetite nanoparticles coated with PEG 3350-Tween 80. Positively charged nanoparticles are more susceptible to being bound and internalized into cancer cells [42]. However, nanoparticles with positively charged surfaces immobilize serum proteins faster than neutrally charged surfaces. This could be a prejudicial propriety because cationic nanoparticles can be loaded with proteins secreted by the immune cells and be recognized and eliminated from blood circulation [43]. Coated nanoparticles obtained from this work shown a low positively charged surface. Positive charges are also correlated with cytotoxicity and hemolysis. That why authors make an emphasis in the evaluation of the interactions of obtained nanoparticles with cells.

3.6. Immunological Characterization of MSC

MSC from adipose tissue are easy to obtain and are available in large quantities, which allows them to be used in different biomedical applications [44]. In this work, the MSC isolated from sheep exhibited fibroblastic morphology with positive expression of CD90 and CD105, which indicates success in the isolation process of MSCs without contamination with other kinds of cells. For the positive control of Anti CD90, a histological section of a sheep brain was used, and the positive control of Anti CD105 was a histological section of a sheep spleen. The positive labeling for the technique was observed through a brown precipitate, which differs from the nuclei in a violet color (Figure 6). CD90 was observed in 88.36% of the cells, while CD105 was observed in the 77.30% of the cells.

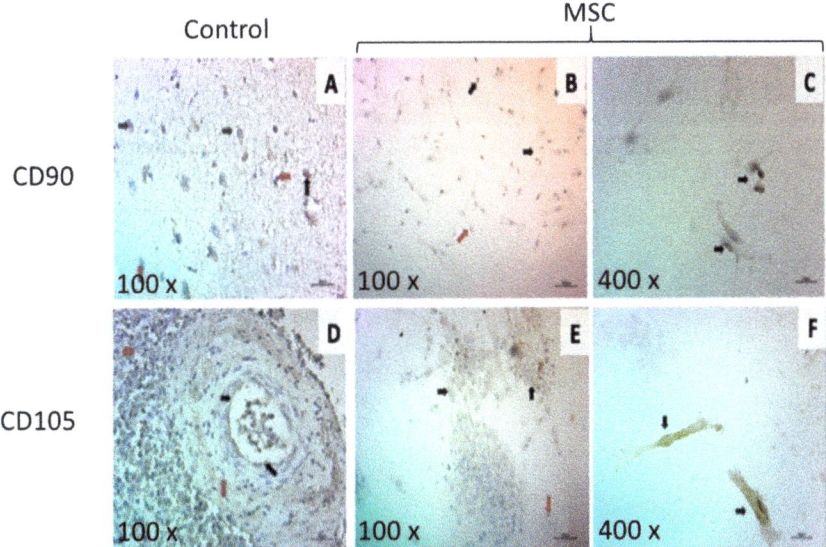

Figure 6. Immunocytochemistry characterization of a mesenchymal stem cell (MSC). (**A**) Ovine brain tissue positive for CD90; (**B**,**C**) MSC marked with CD90; (**D**) ovine spleen tissue positive for CD105 and (**E**,**F**) MSC marked with CD105.

3.7. Cytotoxicity Test

Since in literature there are several findings of harm nanoparticles that damage circulatory, digestive, endocrine, immune, integumentary, nervous, reproductive, respiratory and urinary systems [45], it is a priority to analyze the cytotoxicity of nanoparticles and the kind of harm they can cause. The 3-(4,5-dimethylthiazol-2-yl)-2,5-diphenyltetrazolium bromide (MTT) is a colorimetric technique that evaluates the metabolic activity of cells, specifically the activity of oxidoreductases enzymes found in the mitochondria. Healthy cells that are exposed to an MTT reagent reduce it into formazan crystals, which are insoluble in water and purple color [46].

The results of this study are shown in Figure 7. At all tested concentrations (from 10 to 10,000 μg/mL), bare magnetite nanoparticles showed no statistical difference ($p = 0.05$) in comparison with the negative control. In the magnetite nanoparticles coated with PEG 3350-Tween 80, only concentrations from 10 to 1000 μg/mL showed no significant difference against the negative control ($p = 0.05$); so at these doses, there is no damage in the cell metabolism and the nanoparticles are not considered a dangerous substance. On the other hand, statistical analyses had shown that at concentrations of 10,000 μg/mL, magnetite nanoparticles coated with PEG 3350-Tween 80 displayed a significant difference with respect to the negative control ($p = 0.05$). Thus, it is not safe to use coated nanoparticles at concentrations of 10,000 μg/mL and higher.

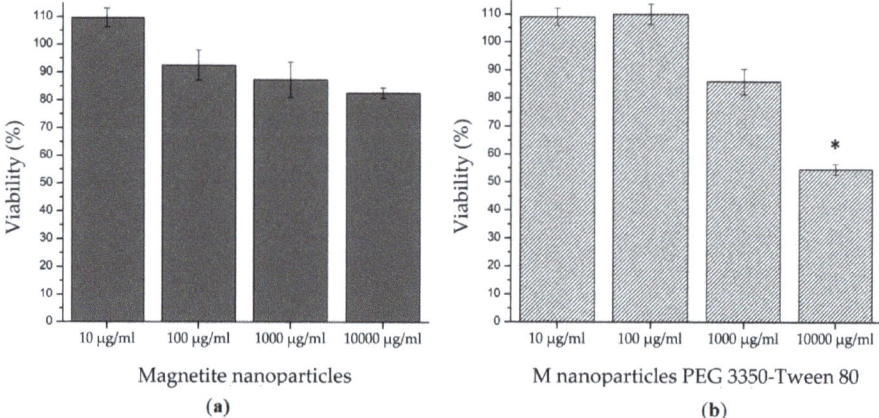

Figure 7. MTT assay results. Both nanoparticles tested, (**a**) bare magnetite nanoparticles and (**b**) magnetite nanoparticles coated with PEG 3350-Tween 80, shown a non-cytotoxic behavior at concentrations from 10 to 1000 µg/mL. At concentrations of 10,000 µg/mL, only magnetite nanoparticles coated with PEG 3350-Tween 80 show a cytotoxic effect with a significant difference* in comparison to the negative control.

3.8. Interactions with Erythrocytes Membranes

Determine if nanoparticles can cause erythrocyte lysis in an ex vivo model is important because when hemolysis occurs in vivo it can cause anemia, jaundice and other pathological conditions. In addition, the hemoglobin released can have a toxic effect on the renal and vascular system. The hemolysis test measures the lysis of the red blood cells exposed to an environmental agent. This lysis produces the release of the intracellular content of the erythrocyte due to the rupture of its membrane. The measured released molecule was the hemoglobin, which is a predominant protein in erythrocytes. According to Standard Practice for Assessment of Hemolytic Properties of Materials, ASTM F756-17, the hemolytic activity of materials is classified in three types: non-hemolytic materials (0%–2% of hemolysis), low hemolytic materials (2%–5% of hemolysis) and high hemolytic materials (higher than 5% of hemolysis) [22]. As is shown in Figure 8 magnetite nanoparticles shown a non-hemolytic behavior at all concentrations tested (from 1 to 1000 µg/mL). On the other hand, magnetite nanoparticles coated with PEG 3350-Tween 80 are safe only at concentrations lower than 100 µg/mL. There was a difference between bare magnetite nanoparticles, which were non-hemolytic at the concentration of 1000 µg/mL, against nanoparticles coated with PEG 3350-Tween 80, which were high hemolytic at the same concentration. This grade of hemotoxicity can be attributed to the positive charges in the coated nanoparticle surface [4].

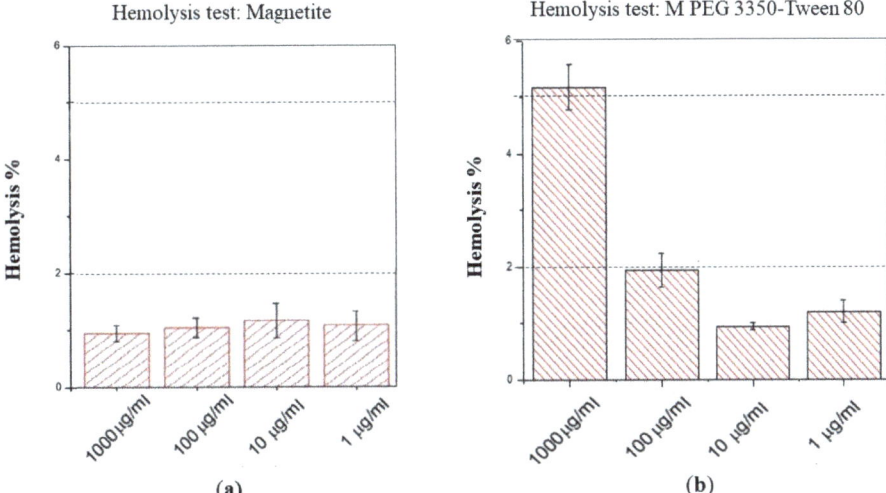

Figure 8. Hemolysis test results. (**a**) Bare magnetite nanoparticles and (**b**) magnetite nanoparticles coated with PEG 3350-Tween 80. Results show that both nanoparticles are considered safe from 1 to 100 µg/mL. Bare magnetite nanoparticles are safe at concentrations of 1000 µg/mL, while magnetite nanoparticles coated with PEG 3350-Tween 80 have shown hemotoxicity at that concentration.

Although there are many advantages of the use of PEG 3350 and Tween 80, such as prolonging the circulation time in the body [12], the correct formulations and concentrations in the use of the coating must be measured and tested because excess of coating can cause several problems at the cellular level. In this work, we found that coating the MNPs did not improve their biological in vitro properties, as observed in the hemolysis and MTT test, but we find that the use of the coat eliminates the presence of agglomerates bigger than 2.7 µm. Some authors findings [47–49] show that PEGylated magnetite nanoparticles are safe to use, but we have to take into consideration that the preparation of the nanoparticle influences in the result of the cytotoxicity tests, as well as the nature of the cell lines or primary cultures employed, the election of the test performed, the PEGylated magnetite doses administrated, the time of exposure and the molecular weight of the PEG employed. Liu et al. in 2017 presented a complete cytotoxic evaluation of different molecular weight PEG molecules. They found that different molecular weight molecules correspond to different cytotoxic behaviors. They also corroborated that at higher doses PEG is more cytotoxic. Liu et al. contributed the PEG cytotoxicity to the reactive oxygen species (ROS) generation, which is generated by PEG derivates [50]. In this study we presented an analysis of the PEG degradation, which showed the disappearance of bonds from the PEG polymer generating derivatives, which correlates also with the formation of ROS. There is no extensive exploration of the formation of ROS and how they affect PEG biological proprieties; a new challenge is opened for researchers who wish to add PEG to their nanostructures for the purpose of improving physical properties of nanoparticles. Another option would be to compare PEG cytotoxicity with different polymers that have given good results in the functionalization of MNPs such as silica [51], chitosan [52] or polyethylenimine [9].

4. Conclusions

In this study, we were able to develop magnetite nanoparticles coated with PEG 3350-Tween 80, as the XRD, EDS and FTIR analyses showed. The SEM and DLS studies showed spherical morphologies and nanometric sizes. The main point of this study was the comparison of the cytotoxicity of bare magnetite nanoparticles compared to magnetite nanoparticles coated with PEG 3350-Tween 80. At low doses of magnetite nanoparticles coated with PEG 3350-Tween 80 are safe to use. The authors recommend the use of primary cell cultures over cell lines to test the cytotoxicity of nanoparticles, due to their sensitivity and similarity to an in vivo environment. Now that authors have developed a non-toxic nanoparticle with the possibility of been targeted to the brain, new in vitro and in vivo testes will be required in order to develop an application.

Author Contributions: Conceptualization, J.A.R.-P., P.E.G.-C. and C.N.S.-D.; methodology, J.A.R.-P., F.G.R.-H., I.A.F.-U., M.B.-V., F.E.P.-M. and E.N.G.-T.; investigation, J.A.R.-P., F.G.R.-H., C.C.-G., P.E.G.-C. and C.N.S.-D.; resources, H.G.M.-R., P.E.G.-C. and C.N.S.-D.; writing—original draft preparation, J.A.R.-P.; writing—review and editing, F.G.R.-H., I.A.F.-U., C.C.-G., H.G.M.-R., F.E.P.-M., P.E.G.-C. and C.N.S.-D.; supervision, C.C.-G., H.G.M.-R., E.N.G.-T., P.E.G.-C. and C.N.S.-D. All authors have read and agreed to the published version of the manuscript.

Funding: This research received no external funding.

Acknowledgments: The authors want to acknowledge the support given by Kassandra O Rodríguez-Aguillón (M.Sc. student, UANL), Alexia Escareño-Cruz (Engr. student, UACJ), Karen S Valenzuela-Villela (M.Sc. student, UACJ), Sergio Lozano Rodríguez (Scientific publications support coordinator, UANL), Hortensia Reyes-Blas (SEM technician, UACJ) and Hugo L Gallardo-Blanco (Sc.D. researcher, UANL).

Conflicts of Interest: The authors declare no conflict of interest.

References

1. Mohammed, L.; Gomaa, H.G.; Ragab, D.; Zhu, J. Magnetic nanoparticles for environmental and biomedical applications: A review. *Particuology* **2017**, *30*, 1–14. [CrossRef]
2. Revia, R.A.; Zhang, M. Magnetite nanoparticles for cancer diagnosis, treatment, and treatment monitoring: Recent advances. *Mater. Today* **2016**, *19*, 157–168. [CrossRef] [PubMed]
3. Lima-Tenório, M.K.; Gómez-Pineda, E.A.; Ahmad, N.M.; Fessi, H.; Elaissari, A. Magnetic nanoparticles: In vivo cancer diagnosis and therapy. *Int. J. Pharm.* **2015**, *493*, 313–327. [CrossRef] [PubMed]
4. Roacho-Perez, J.A.; Gallardo-Blanco, H.L.; Sanchez-Dominguez, M.; Garcia-Casillas, P.E.; Chapa-Gonzalez, C.; Sanchez-Dominguez, C.N. Nanoparticles for death-induced gene therapy in cancer (Review). *Mol. Med. Rep.* **2017**, *17*, 1413–1420. [CrossRef] [PubMed]
5. Múzquiz-Ramos, E.M.; Guerrero-Chávez, V.; Macías-Martínez, B.I.; López-Badillo, C.M.; García-Cerda, L.A. Synthesis and characterization of maghemite nanoparticles for hyperthermia applications. *Ceram. Int.* **2014**, *41*, 397–402. [CrossRef]
6. Karimi, Z.; Karimi, L.; Shokrollahi, H. Nano-magnetic particles used in biomedicine: Core and coating materials. *Mater. Sci. Eng. C* **2013**, *33*, 2465–2475. [CrossRef]
7. Rivera-González, H.; Torres-Pacheco, L.; Álvarez-Contreras, L.; Olivas, A.; Guerra-Balcázar, M.; Valdez, R.; Arjona, N. Synthesis of Pd–Fe$_3$O$_4$ nanoparticles varying the stabilizing agent and additive and their effect on the ethanol electro-oxidation in alkaline media. *J. Electroanal. Chem.* **2019**, *835*, 301–312. [CrossRef]
8. Ríos-Hurtado, J.C.; Martínez-Valdés, A.C.; Rangel-Méndez, J.R.; Ballesteros-Pacheco, J.C.; Múzquiz-Ramos, E.M. Facile synthesis and characterization of Mn$_x$Zn$_{1-x}$Fe$_2$O$_4$/activated carbon composites for biomedical applications. *J. Ceram. Sci. Technol.* **2016**, *7*, 289–294. [CrossRef]
9. Megías, R.; Arco, M.; Ciriza, J.; Saenz-Burgo, L.; Puras, G.; López-Viota, M.; Delgado, A.V.; Dobson, J.P.; Arias, J.L.; Pedraz, J.L. Design and characterization of a magnetite/PEI multifunctional nanohybrid as non-viral vector and cell isolation system. *Int. J. Pharm.* **2017**, *518*, 270–280. [CrossRef]
10. Veronese, F.M.; Mero, A. The impact of PEGylation on biomedical therapies. *BioDrugs* **2008**, *22*, 315–329. [CrossRef]
11. Turecek, P.L.; Bossard, M.J.; Schoetens, F.; Ivens, I.A. PEGylation of Biopharmaceuticals: A Review of Chemistry and Nonclinical Safety Information of Approved Drugs. *J. Pharm. Sci.* **2016**, *105*, 460–475. [CrossRef] [PubMed]
12. Suk, J.S.; Xu, Q.; Kim, N.; Hanes, J.; Ensign, L.M. PEGylation as a strategy for improving nanoparticle-based drug and gene delivery. *Adv. Drug Deliv. Rev.* **2016**, *99*, 28–51. [CrossRef] [PubMed]
13. Li, X.; Meng, F.; Zhang, S.; Zhang, M.; Zeng, Y.; Zhong, Q. The effect of polyethylene glycol modification on CrO$_x$/TiO$_2$ catalysts for NO oxidation. *Colloids Surfaces A Physicochem.* **2019**, *578*, 123588. [CrossRef]
14. Favela-Camacho, S.E.; Pérez-Robles, J.F.; García-Casillas, P.E.; Godinez-Garcia, A. Stability of magnetite nanoparticles with different coatings in a simulated blood plasma. *J. Nanoparticle Res.* **2016**, *18*, 176. [CrossRef]

15. Yadav, M.; Parle, M.; Sharma, N.; Dhingra, S.; Raina, N.; Jindal, D.K. Brain targeted oral delivery of doxycycline hydrochloride encapsulated Tween 80 coated chitosan nanoparticles against ketamine induced psychosis: Behavioral, biochemical, neurochemical and histological alterations in mice. *Drug Deliv.* **2017**, *24*, 1429–1440. [CrossRef] [PubMed]
16. Ma, Y.; Zheng, Y.; Zeng, X.; Jiang, L.; Chen, H.; Liu, R.; Huang, L.; Mei, L. Novel docetaxel-loaded nanoparticles based on PCL-Tween 80 copolymer for cancer treatment. *Int. J. Nanomed.* **2011**, *6*, 2679–2688. [CrossRef]
17. Yuan, X.; Ji, W.; Chen, S.; Bao, Y.; Tan, S.; Lu, S.; Wu, K.; Chu, Q. A novel paclitaxel-loaded poly(D,L-lactide-co-glycolide)-Tween 80 copolymer nanoparticle overcoming multidrug for lung cancer treatment. *Int. J. Nanomed.* **2016**, *11*, 2119–2131. [CrossRef]
18. Hussain, M.Z.; Khan, R.; Ali, R.; Khan, Y. Optical properties of laser ablated ZnO nanoparticles prepared with Tween-80. *Mater. Lett.* **2014**, *122*, 147–150. [CrossRef]
19. Khan, Y.; Durrani, S.K.; Siddique, M.; Mehmood, M. Hydrothermal synthesis of alpha Fe_2O_3 nanoparticles capped by Tween-80. *Mater. Lett.* **2011**, *65*, 2224–2227. [CrossRef]
20. Chapa-Gonzalez, C.; Roacho-Pérez, J.A.; Martínez-Pérez, C.A.; Olivas-Armendáriz, I.; Jimenez-Vega, F.; Castrejon-Parga, K.Y.; García-Casillas, P.E. Surface modified superparamagnetic nanoparticles: Interaction with fibroblasts in primary cell culture. *J. Alloys Compd.* **2015**, *615*, S655–S659. [CrossRef]
21. Bhattacharjee, S. DLS and zeta potential—What they are and what they are not? *J. Control. Release* **2016**, *235*, 337–351. [CrossRef] [PubMed]
22. Macías-Martínez, B.I.; Cortés-Hernández, D.A.; Zugasti-Cruz, A.; Cruz-Ortíz, B.R.; Múzquiz-Ramos, E.M. Heating ability and hemolysis test of magnetite nanoparticles obtained by a simple co-precipitation method. *J. Appl. Res. Technol.* **2016**, *14*, 239–244. [CrossRef]
23. Anbarasu, M.; Anandan, M.; Chinnasamy, E.; Gopinath, V.; Balamurugan, K. Synthesis and characterization of polyethylene glycol (PEG) coated Fe_3O_4 nanoparticles by chemical co-precipitation method for biomedical applications. *Spectrochim. Acta-Part A Mol. Biomol. Spectrosc.* **2015**, *135*, 536–539. [CrossRef]
24. Aghazadeh, M.; Ganjali, M.R. One-step electro-synthesis of Ni^{2+} doped magnetite nanoparticles and study of their supercapacitive and superparamagnetic behaviors. *J. Mater. Sci. Mater. Electron.* **2018**, *29*, 4981–4991. [CrossRef]
25. Habibi, N. Preparation of biocompatible magnetite-carboxymethyl cellulose nanocomposite: Characterization of nanocomposite by FTIR, XRD, FESEM and TEM. *Spectrochim. Acta-Part A Mol. Biomol. Spectrosc.* **2014**, *131*, 55–58. [CrossRef] [PubMed]
26. Gao, M.; Li, W.; Dong, J.; Zhang, Z.; Yang, B. Synthesis and Characterization of Superparamagnetic Fe_3O_4@SiO_2 Core-Shell Composite Nanoparticles. *World J. Condens. Matter Phys.* **2011**, *1*, 49–54. [CrossRef]
27. Sarkar, T.; Rawat, K.; Bohidar, H.B.; Solanki, P.R. Electrochemical immunosensor based on PEG capped iron oxide nanoparticles. *J. Electroanal. Chem.* **2016**, *783*, 208–216. [CrossRef]
28. Lal, M.; Verma, S.R. Synthesis and Characterization of Poly Vinyl Alcohol Functionalized Iron Oxide Nanoparticles. *Macromol. Symp.* **2017**, *376*, 1700017. [CrossRef]
29. Khalil, M.I. Co-precipitation in aqueous solution synthesis of magnetite nanoparticles using iron(III) salts as precursors. *Arab. J. Chem.* **2015**, *8*, 279–284. [CrossRef]
30. Mata-Pérez, F.; Martínez, J.R.; Guerrero, A.L.; Ortega-Zarzosa, G. New way to produce magnetite nanoparticles at low temperature. *Adv. Chem. Eng. Res.* **2015**, *4*. [CrossRef]
31. Khanna, L.; Verma, N.K. Study on novel, superparamagnetic and biocompatible $PEG/KFeO_2$ nanocomposite. *J. Appl. Biomed.* **2015**, *13*, 23–32. [CrossRef]
32. Zargar, T.; Kermanpur, A.; Labbaf, S.; Houreh, A.B.; Esfahani, M.H.N. PEG coated $Zn_{0.3}Fe_{2.7}O_4$ nanoparticles in the presence of <alpha>Fe_2O_3 phase synthesized by citric acid assisted hydrothermal reduction process for magnetic hyperthermia applications. *Mater. Chem. Phys.* **2018**, *212*, 432–439. [CrossRef]
33. Wu, X.; Wang, W.; Li, F.; Khaimanov, S.; Tsidaeva, N.; Lahoubi, M. PEG-assisted hydrothermal synthesis of $CoFe_2O_4$ nanoparticles with enhanced selective adsorption properties for different dyes. *Appl. Surf. Sci.* **2016**, *389*, 1003–1011. [CrossRef]
34. Fu, X.; Kong, W.; Zhang, Y.; Jiang, L.; Wang, J.; Lei, J. Novel solid-solid phase change materials with biodegradable trihydroxy surfactants for thermal energy storage. *RSC Adv.* **2015**, *5*, 68881–68889. [CrossRef]
35. Atabaev, T.S. PEG-Coated superparamagnetic dysprosium-doped Fe_3O_4 nanoparticles for potential MRI imaging. *Bio Nano Sci.* **2018**, *8*, 299–303. [CrossRef]

36. Tudisco, C.; Bertani, F.; Cambria, M.T.; Sinatra, F.; Fantechi, E.; Inoocenti, C.; Sangregorio, C.; Dalcanale, E.; Condorelli, G.G. Functionalization of PEGylated Fe$_3$O$_4$ magnetic nanoparticles with tetraphosphonate cavitand for biomedical application. *Nanoscale* **2013**, *5*, 11438–11446. [CrossRef]
37. Quevedo, I.R.; Olsson, A.L.J.; Clark, R.J.; Veinot, J.G.C.; Tufenkji, N. Interpreting deposition behavior of polydisperse surface-modified nanoparticles using QCM-D and sand-packed columns. *Environ. Eng. Sci.* **2014**, *31*. [CrossRef]
38. Champion, J.A.; Mitragotri, S. Role of target geometry in phagocytosis. *Proc. Natl. Acad. Sci. USA* **2006**, *103*, 4930–4934. [CrossRef]
39. Geng, Y.; Dalhaimer, P.; Caí, S.; Tsai, R.; Tewari, M.; Minko, T.; Discher, D.E. Shape effects of filaments versus spherical particles in flow and drug delivery. *Nat. Nanotechnol.* **2007**, *2*, 249–255. [CrossRef]
40. Cabral, H.; Matsumoto, Y.; Mizuno, K.; Chen, Q.; Murakami, M.; Kimura, M.; Terada, Y.; Kano, M.R.; Miyazono, K.; Uesaka, M.; et al. Accumulation of sub-100 nm polymeric micelles in poorly permeable tumors depends on size. *Nat. Nanotechnol.* **2011**, *6*, 815–823. [CrossRef]
41. Soo-Choi, H.; Liu, W.; Misra, P.; Tanaka, E.; Zimmer, J.P.; Ipe, B.I.; Bawendi, M.G.; Frangioni, J.V. Renal clearance of quantum dots. *Nat. Biotechnol.* **2007**, *25*, 1165–1170. [CrossRef] [PubMed]
42. Thurston, G.; McLean, J.W.; Rizen, M.; Baluk, P.; Haskell, A.; Murphy, T.J.; Hanahan, D.; McDonald, D.M. Cationic liposomes target angiogenic endothelial cells in tumors and chronic inflammation in mice. *J. Clin. Investig.* **1998**, *101*, 1401–1413. [CrossRef] [PubMed]
43. Roser, M.; Fischer, D.; Kissel, T. Surface-modified biodegradable albumin nano and microspheres: Effect of surface charges on in vitro phagocytosis and biodistribution in rats. *Eur. J. Pharm. Biopharm.* **1998**, *46*, 255–263. [CrossRef]
44. De Girolamo, L.; Lucarelli, E.; Alessandri, G.; Avanzini, M.A.; Bernardo, M.E.; Biagi, E.; Brini, A.T.; D'Amico, G.; Fagioli, F.; Ferrero, I.; et al. Mesenchymal Stem/Stromal Cells: A New "Cells as Drugs" Paradigm. Efficacy and Critical Aspects in Cell Therapy. *Curr. Pharm. Des.* **2013**, *19*, 2459–2473. [CrossRef] [PubMed]
45. Jiang, Z.; Shan, K.; Song, J.; Liu, J.; Rajendran, S.; Pugazhendhi, A.; Jacob, J.A.; Chen, B. Toxic effects of magnetic nanoparticles on normal cells and organs. *Life Sci.* **2019**, *220*, 156–161. [CrossRef]
46. Gomez-Perez, M.; Fourcade, L.; Mateescu, M.A.; Paquin, J. Neutral Red versus MTT assay of cell viability in the presence of copper compounds. *Anal. Biochem.* **2017**, *535*, 43–46. [CrossRef]
47. Ayubi, M.; Karimi, M.; Adbpour, S.; Rostamizadeh, K.; Parsa, M.; Zamani, M.; Saedi, A. Magnetic nanoparticles decorated with PEGylated curcumin as dual targeted drug delivery: Synthesis, toxicity and biocompatibility study. *Mater. Sci. Eng. C* **2019**, *104*, 109810. [CrossRef]
48. Hu, J.; Obayemi, J.D.; Malatesta, K.; Košmrlj, A.; Soboyejo, W.O. Enhanced cellular uptake of LHRH-conjugated PEG-coated magnetite nanoparticles for specific targeting of triple negative breast cancer cells. *Mater. Sci. Eng. C* **2018**, *88*, 32–45. [CrossRef]
49. Illés, E.; Szekerez, M.; Kupcsik, E.; Tóth, I.Y.; Farkas, K.; Jedlovszky-Hajdú, A.; Tombácz, E. PEGylation of surfaced magnetite core-shell nanoparticles for biomedical application. *Colloids Surfaces A Physicochem. Eng. Asp.* **2014**, *460*, 429–440. [CrossRef]
50. Liu, G.; Li, Y.; Yang, L.; Wei, Y.; Wang, X.; Wang, Z.; Tao, L. Cytotoxicity study of polyethylene glycol derivates. *RSC Adv.* **2017**, *7*, 18252–18259. [CrossRef]
51. Candian-Lobato, N.C.; Mello-Ferreira, A.; Weidler, P.G.; Franzreb, M.; Borges-Mansur, M. Improvement of magnetic solvent extraction using functionalized silica-coated Fe$_3$O$_4$ nanoparticles. *Sep. Purif. Technol.* **2019**, *229*, 115839. [CrossRef]
52. Pourmortazavi, S.M.; Sahebi, H.; Zandavar, H.; Mirsadeghi, S. Fabrication of Fe$_3$O$_4$ nanoparticles coated by extracted shrimp peels chitosan as sustainable adsorbents for removal of chromium contaminates from wastewater: The design of experiment. *Compos. B Eng.* **2019**, *175*, 107130. [CrossRef]

© 2020 by the authors. Licensee MDPI, Basel, Switzerland. This article is an open access article distributed under the terms and conditions of the Creative Commons Attribution (CC BY) license (http://creativecommons.org/licenses/by/4.0/).

Article

Non-Chloride in Situ Preparation of Nano-Cuprous Oxide and Its Effect on Heat Resistance and Combustion Properties of Calcium Alginate

Peiyuan Shao [1], Peng Xu [1], Lei Zhang [1,2], Yun Xue [1], Xihui Zhao [1], Zichao Li [1,2,*] and Qun Li [1,3,*]

[1] College of Chemistry and Chemical Engineering, Qingdao University, Qingdao 266071, China; 15666681178@163.com (P.S.); X1627271005@163.com (P.X.); 2017020832@qdu.edu.cn (L.Z.); xueyun@qdu.edu.cn (Y.X.); zhaoxihui@qdu.edu.cn (X.Z.)
[2] College of Life Sciences, Institute of Advanced Cross-Field Science, Qingdao University, Qingdao 266071, China
[3] Shandong Collaborative Innovation Center of Marine Biobased Fibers and Ecological Textiles, Qingdao University, Qingdao 266071, China
* Correspondence: zichaoli@qdu.edu.cn (Z.L.); qunli@qdu.edu.cn (Q.L.); Tel.: +86-532-8595-0705 (Q.L.)

Received: 11 October 2019; Accepted: 24 October 2019; Published: 27 October 2019

Abstract: With Cu^{2+} complexes as precursors, nano-cuprous oxide was prepared on a sodium alginate template excluded of Cl^- and based on which the calcium alginate/nano-cuprous oxide hybrid materials were prepared by a Ca^{2+} crosslinking and freeze-drying process. The thermal degradation and combustion behavior of the materials were studied by related characterization techniques using pure calcium alginate as a comparison. The results show that the weight loss rate, heat release rate, peak heat release rate, total heat release rate and specific extinction area of the hybrid materials were remarkably lower than pure calcium alginate, and the flame-retardant performance was significantly improved. The experimental data indicates that nano-cuprous oxide formed a dense protective layer of copper oxide, calcium carbonate and carbon by lowering the initial degradation temperature of the polysaccharide chain during thermal degradation and catalytically dehydrating to char in the combustion process, and thereby can isolate combustible gases, increase carbon residual rates, and notably reduce heat release and smoke evacuation.

Keywords: alginate; non-chloride in situ preparation; nano-cuprous oxide; flame-retardant; mechanism

1. Introduction

Nano-cuprous oxide is a typical p-type semiconductor material with active electron holes, which exhibits effective catalytic activity, strong adsorption and low-temperature paramagnetic properties [1]. It has potential applications in organic synthesis, photoelectric conversion, new energy, water decomposition, sterilization, superconductivity, and others, and has become one of the most widely studied inorganic nanomaterials in the past decade [2–10].

Due to their low cost and sustainability, marine biomaterials from renewable resources have attracted attention today as fossil energy is depleting. Alginate is a linear, binary-block copolymer with special structures and properties [11,12]. It has served as a template for the preparation of new functional hybrid materials in recent years [13–17]. In particular, sodium alginate can form a strong and stable "egg box" model (Figure 1) [12,18,19] structure of water-insoluble materials by crosslinking with divalent or trivalent metal ions, which is not only widely used in food and medicine [20–22], but also as an eco-friendly flame-retardant material [23–27].

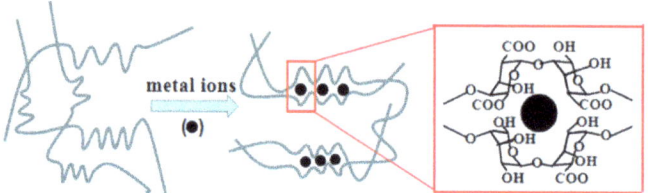

Figure 1. The crosslinking of sodium alginate—the "egg box" structure.

It is well known that the goal of developing hybrid materials is to improve the functionality of the hybrid material by creating a synergistic effect through the interaction between two phases [13,28–30], and the flame-retardant properties of the inherently flame-retardant calcium alginate can be further improved by adding inorganic nanoparticles [31–35]. In order to overcome the defects caused by the agglomeration of inorganic nanoparticles due to the mechanical addition method [34], previous research on preparing calcium alginate-based hybrid materials by in situ methods have been reported [31–33]. Herein, a new multifunctional material employing nano-cuprous oxide to improve the antibacterial and catalytic properties of calcium alginate is presented. However, to the best of our knowledge, no such report has been well documented so far. Note that copper alginate shows no flame retardancy [36]; thus, the effect of nano-cuprous oxide on the flame-retardant properties of calcium alginate is worth exploring.

Generally, calcium chloride has been used as a crosslinking agent to prepare calcium alginate. However, since copper ions, cuprous ions, and chloride ions have complicated chemical reactions, it is difficult to prepare calcium alginate/nano-cuprous oxide hybrid materials by such a method based on the attempts in our laboratory for over a year. Therefore, we propose a novel "non-chlorine in situ" preparation method, aiming for the facile preparation of calcium alginate/nano-cuprous oxide hybrid materials by the in situ method in the absence of chloride ions. Our study found an improved limiting oxygen index of 54.0 for the hybrid material compared to pure calcium alginate. Although the role of inorganic nanoparticles in hybrid material has been reported so far, there has been no literature on the effects of nano-cuprous oxide on the flame-retardant properties of calcium alginate or its mechanism. The results here show that the peak heat release rate of calcium alginate can be reduced by 15% with only 1 wt % nano-cuprous oxide addition, and the flame-retardant mechanism was slightly different from the traditional mechanism [5,7]. In summary, the calcium alginate/nano-cuprous oxide hybrid materials prepared by the "non-chlorine in situ" method is a multi-functional material with important potential applications and flame-retardant, antibacterial, and catalytic properties. The effect and mechanism of cuprous oxide on the material is essential to the design of the material.

2. Materials and Methods

2.1. Materials

Sodium alginate (NaAlg) was purchased from the Guangfu Fine Chemical Research Institute (Tianjin, China). Pentahydrate copper sulfate ($CuSO_4 \cdot 5H_2O$) was obtained from the Guangcheng Chemical Reagent Co., Ltd. (Tianjin, China). Both ammonia water ($NH_3 \cdot H_2O$) and sodium hydroxide (NaOH) were supplied by the fine chemical plant of the Laiyang Economic and Technological Development Zone (Yantai, China). Calcium acetate and ascorbic acid were purchased from Sinopharm Chemical Reagent Co., Ltd. (Shanghai, China). All the chemicals were of analytical grade and not further purified, and all the solutions were prepared using deionized water.

2.2. Preparation of Nano-Cu_2O In Situ

The preparation process of the nano-Cu_2O is shown schematically in Scheme 1. $CuSO_4 \cdot 5H_2O$ (3.9 g) was dissolved in a predetermined amount of deionized water, followed by the addition of 13 mL $NH_3 \cdot H_2O$ (12 mol/L). Then, 20 g of NaAlg was added and the whole volume of the mixture solution was maintained at 500 mL. The solution was heated at a constant temperature of 55 °C for 6 h until a uniform and stable blue solution was formed. Subsequently, 11.4 g of ascorbic acid was added into the solution and was slowly stirred until the color entirely changed to yellow. Consequently, the nano-Cu_2O was prepared on the NaAlg template.

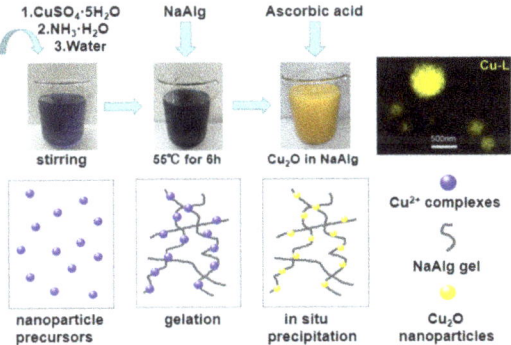

Scheme 1. Schematic of nano-Cu_2O preparation in situ.

2.3. Preparation of Calcium Alginate/Nano-Cuprous Oxide Hybrid Materials

NaOH (1 mol/L) was slowly dropped into the 3 wt % calcium acetate solution (2 L) until the pH reached 10. The as-prepared solution containing 4 wt % NaAlg was poured into molds of various shapes and then placed in the calcium acetate solution for 24 h, obtaining the crosslinked hybrid material (CaAlg/nano-Cu_2O). Finally, the product was washed several times with deionized water and dried in a freeze-dryer for 24 h. It is found by calculation that the content of Cu_2O in the hybrid material prepared by the in situ method was equivalent to the mechanical process of adding of 1 wt % Cu_2O. As a comparison, pure calcium alginate (CaAlg) was prepared in a similar procedure only without the addition of the reagents related to the preparation of Cu_2O.

2.4. Measurements and Characterizations

For the characterization of the calcined residue, the samples were heated to a target temperature (250, 450, and 750 °C) at a heating rate of 5 °C/min in air, and then kept for 2 h.

2.4.1. Density

The quality and dimension of CaAlg and CaAlg/nano-Cu_2O hybrid material were measured using an electronic balance (PL303 Precision Balance, Mettler-Toledo, Tokyo, Japan) and digital calipers, and then the densities were calculated based on the data.

2.4.2. X-ray Diffraction (XRD)

An X-ray diffractometer (Ultima IV, Rigaku, Tokyo, Japan) was employed to investigate the samples and calcined residue under Cu Kα radiation (λ = 0.15418 nm), with a scan rate (2θ) of 5°/min in an angular range of 15–80 in continuous mode (operating voltage: 40 kV, current: 30 mA).

2.4.3. Field Emission Scanning Electron Microscopy (FESEM)

The morphology and microstructure of the materials and their calcined residues were observed by a field emission scanning electron microscope (Sigma 500/VP, Carl Zeiss AG, Jena, Germany) at a beam voltage of 10 kV. Prior to SEM imaging, all the sample surfaces were sprayed with thin layers of gold to avoid charging under the electron beam. The distribution of Cu on the surface of the CaAlg/nano-Cu_2O hybrid material was plotted using an energy dispersive X-ray spectrometer (EDX) attachment.

2.4.4. Synchronous Thermal Analyzer (STA)

A synchronous thermal analyzer (TGA/DSC1 1600, Mettler Toledo, Columbus, OH, USA) was employed to analyze the materials under a nitrogen atmosphere. Samples of 4–6 mg were placed in an aluminum crucible and heated from room temperature to 1000 °C at a heating rate of 10 °C/min and a gas flow rate of 50 mL/min.

2.4.5. Cone Calorimeter (CONE)

The combustion behavior of the samples was studied with a cone calorimeter (FTT-0007, Fire Testing Technology, East Grinstead, UK) under an external heat flux of 50 kW/m^2 with horizontal orientation in air according to standard ISO 5660-1. The dimensions of all the samples were 100 mm × 100 mm × 8 mm.

2.4.6. Limiting Oxygen Index (LOI)

The LOI values were determined using a M606B type instrument (Shanfang Instrument Co., Ltd., Qingdao, China). Samples of dimensions 80 mm × 10 mm × 8 mm were used for all the tests while CaAlg/nano-Cu$_2$O was compressed to ensure that it was substantially the same quality as CaAlg. The samples were placed in a standard laboratory, at a temperature of 23 ± 2 °C and a relative humidity of 50 ± 5%, for at least 24 h before testing.

2.4.7. Vertical Burning (UL-94)

Vertical burning tests were performed on a CZF-2 test instrument (Jiangning Analytical Instrument Factory, Nanjing, China) according to the standard American Society for Testing and Materials (ASTM) D3801. The size of both samples was 130 mm × 13 mm × 8 mm.

2.4.8. Microscale Combustion Calorimeter (MCC)

A microscale combustion calorimeter (MCC-2, Govmark, Farmingdale, NY, USA) was used to investigate the flammability behavior of the test samples. Samples of 5 ± 0.05 mg were heated from room temperature to 750 °C (heating rate of 1 °C/s) in a stream of nitrogen flowing at 80 cm^3/min. Before entering a 900 °C combustion furnace, the thermal degradation products were mixed with a 20 cm^3/min steam, consisting of 20% oxygen and 80% nitrogen.

3. Results and Discussion

3.1. Morphology of Materials

In Figure 2a, the digital photo shows that the two materials exhibited similar porous structures with notably different colors. The difference in color can be attributed to the fact that nano-Cu$_2$O changed the optical properties of the material itself, which was dependent on the structure, size, and dispersion of the particles [13]. The freeze-drying condensed the moisture in the material into ice, and the "ice matrix" left after the sublimation of the ice was the essence of its macroscopic appearance as a porous structure [37,38].

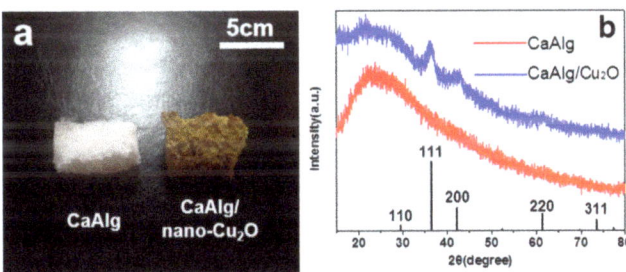

Figure 2. Digital photos (**a**) and X-ray diffraction (XRD) patterns (**b**) of CaAlg and CaAlg/nano-Cu_2O.

The two materials were analyzed by XRD and the results are shown in Figure 2b. Comparing the obtained diffraction peak with the pure phase Cu_2O (cuprite, JCPDS-#99-0041), it was found that the four peaks with different diffraction values of two materials (at 36.4°, 42.3°, 61.4° and 73.5°) were identical with the crystal planes of Cu_2O at (111), (200), (220), and (311), respectively. This confirms the existence of Cu_2O and also the formation of hybrid materials.

It is noteworthy that the density of CaAlg was 0.19 g/cm^3 while that of CaAlg/nano-Cu_2O hybrid material was only 0.08 g/cm^3. FESEM was used to study the morphology of the two materials in order to explain the difference in density. The results of electron microscopic characterization are shown in Figure 3. It can be seen from the analysis and comparison of the SEM images that CaAlg showed a lamellar structure with a smooth, uniform surface and a few nodules. CaAlg/nano-Cu_2O was a three-dimensional network structure, and some spherical Cu_2O particles with an average diameter below 100 nm were adhered to the CaAlg framework interface through a strong interaction, while the others were embedded in the framework of CaAlg.

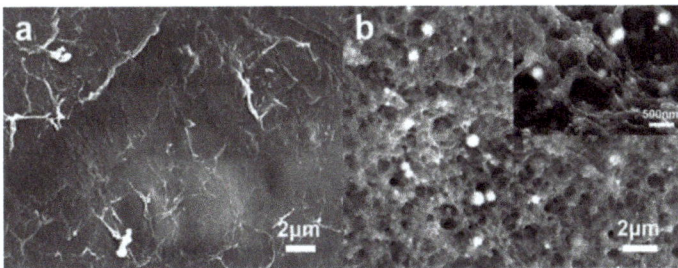

Figure 3. Scanning electron microscopy (SEM) images of (**a**) CaAlg and (**b**) CaAlg/nano-Cu_2O. The inset in (**b**) is the magnified image.

It can be noticed from the microstructure that there were differing densities between the two materials and the three-dimensional network structure clearly exhibited a relatively lower density. The formation of this type of structure is due to the fact that NaAlg itself is rich in hydroxyl and carboxyl groups and can form intermolecular hydrogen bonds with Cu_2O in situ [39,40]. The continuous network structure formed at the same time affected the mass transfer process of the solid–liquid phases in subsequent crosslinking, which led to the presence of a similar three-dimensional network structure in the final hybrid materials. This structure can effectively prevent the movement of polymer molecular chains during combustion with a lower thermal conductivity, and thus can provide better thermal stability of the materials [11,11].

3.2. Thermal Stability

The thermal stability of CaAlg and CaAlg/nano-Cu$_2$O was investigated by simultaneous thermal analysis (TG-DSC) and the comparison curves are displayed in Figure 4. The studied samples followed the same degradation trend and tended to be stable after three stages of decline. The decomposition process of CaAlg was similar to that of previous studies [31,42], and the three stages of its decomposition included: the desorption process of water at 50–140 °C, the thermal destruction of glycosidic bonds and the elimination process of adjacent hydroxyl groups at 240–400 °C; and a process took place in which the glycosidic bond was further destroyed to form an intermediate substance at 400–520 °C.

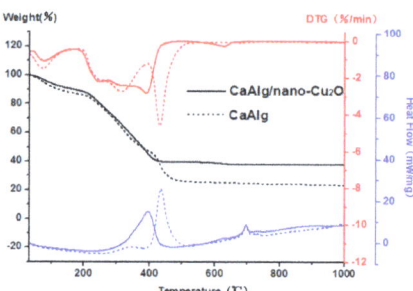

Figure 4. Weight loss, DTG, and DSC curves of CaAlg and CaAlg/nano-Cu$_2$O under N$_2$.

Comparing the two materials, the thermal stability of CaAlg/nano-Cu$_2$O was obviously better than that of CaAlg, and the rate of weight loss was reduced by about 15%. In the desorption stage of water, the weight loss rate of CaAlg/nano-Cu$_2$O was significantly lower than that of CaAlg, which can be attributed to the low thermal conductivity for the three-dimensional network structure of the hybrid material. As a biogenic polymer, CaAlg is rich in hydroxyl and carboxyl groups and can easily absorb water, while for hybrid materials, Cu$_2$O can form multi-hydrogen bonds with the hydroxyl and carboxyl groups in CaAlg, so that the composites had a better thermal stability at low temperature.

As shown in Figure 4, more notable differences occurred at around 400 °C, in which CaAlg/nano-Cu$_2$O was the first to re-decompose the glycosidic bonds at 360 °C, almost 40 °C less than CaAlg. This was mainly due to the fact that Cu$_2$O in the hybrid material catalyzed the thermal decomposition of CaAlg and reduced the energy required for the reaction [43]. Comparing the DSC curves, it was found that although the decomposition reaction of CaAlg/nano-Cu$_2$O proceeded in advance, the peak exotherm was significantly lower than that of CaAlg (namely 38.8%). The decrease in peak heat release indicated improvement in the thermal stability of the material.

In addition, there were two other noticeable observations in the curves shown in Figure 4. The slight weight loss of CaAlg/nano-Cu$_2$O at 600 °C led to a weak change in the heat flux, which may be due to the decomposition of CaCO$_3$ in the thermal cracking products of CaAlg catalyzed by CuO. It can be observed from the DSC curves that the two materials exhibited exothermic peaks at 700 °C without causing mass loss, which can be attributed to the phase transition process of the intermediate produced by the thermal decomposition of CaAlg at high temperature.

The XRD analysis of the calcined residue of the two prepared materials was carried out and the results are shown in Figure 5. It can be seen that CaAlg/nano-Cu$_2$O showed a diffraction peak of Cu$_2$O (cuprite, JCPDS-#99-0041) only in the calcination residue at 250 °C. However, in the calcination residue under the three temperature conditions, clear CuO (copper oxide, JCPDS-#89-5898) diffraction peaks can be observed, which were formed by the oxidation of Cu$_2$O in the material. Except for Cu$_2$O and CuO, the components of the calcined residue of the two materials were basically the same.

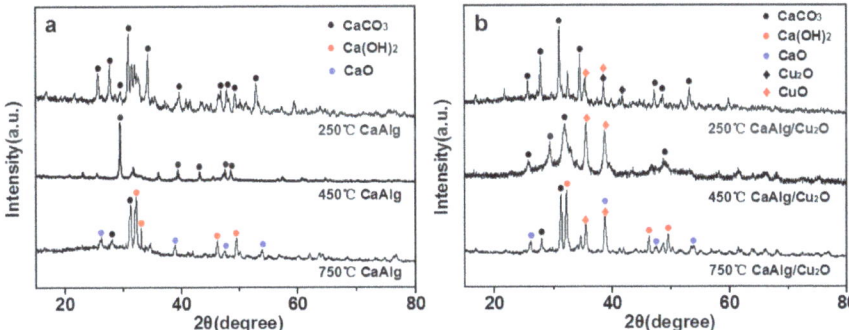

Figure 5. XRD patterns of calcined residue at different temperatures: (**a**) CaAlg and (**b**) CaAlg/nano-Cu$_2$O hybrid material.

Among them, one category of product was CaCO$_3$ of different crystal phases at 250 °C, including calcite (calcite, JCPDS-#99-0022), aragonite (aragonite, JCPDS-#71-2392) and vaterite (vaterite, JCPDS-#74-1867), while the product was at a single calcite phase of CaCO$_3$ at 450 °C. With the decomposition of CaCO$_3$ at 750 °C, not only CaCO$_3$ but also CaO and Ca(OH)$_2$ were generated.

Through the morphological study of the residue after calcination of the two materials, the stability of the materials was analyzed from a microscopic point of view. The relevant characterization results are shown in Figure 6.

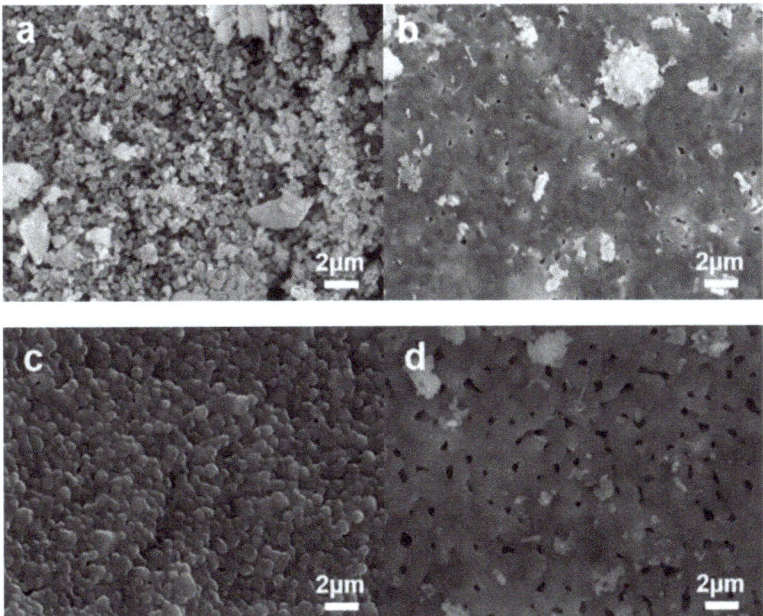

Figure 6. SEM images of calcined residue: (**a**) CaAlg heated to 250 °C; (**b**) CaAlg/nano-Cu$_2$O heated to 250 °C; (**c**) CaAlg heated to 750 °C; (**d**) CaAlg/nano-Cu$_2$O heated to 750 °C.

As shown in Figure 6a, CaCO$_3$ particles of different sizes were distributed on the surface of the CaAlg calcination residue at 250 °C. However, for CaAlg/nano-Cu$_2$O, the framework of CaAlg collapsed as the temperature rose and Cu$_2$O particles with a higher surface energy were rapidly deposited on the surface of the material, which was oxidized to CuO. As seen in Figure 6b, Cu_2O catalyzed the decomposition reaction of the outer layer of CaAlg, in which the release of gas carried off most of the heat and the remaining solid residue formed a dense barrier on the outer layer. The formation of the barrier layer can prevent heat and combustible gas from diffusing into the internal layer [44], greatly reducing the heat release rate and significantly improving the flame retardancy.

It can be observed from Figure 6c that dense polyhedral particles covered the surface of the CaAlg calcination residue at 750 °C, demonstrating the flame-retardant properties of the material. However, the most obvious morphological change in the calcination residue of CaAlg/nano-Cu_2O was the inflation of the residue particles [45], as displayed in Figure 6d, which directly formed denser particle distribution and smaller surface voids. This can further improve the flame retardancy of the material. The morphological difference was related to the carbonized layer formation of the CuO-promoted material at high temperature [11,28].

3.3. Combustion Behavior

The experimental data of cone calorimetry are listed in Table 1. As seen, the heat release rate (HRR), peak heat release rate (PHRR), and total heat of release (THR) of CaAlg/nano-Cu_2O were significantly lower than those of CaAlg. However, the quality of the samples used in this test was quite different due to the various structures, making the comparison of data inconclusive. At present, there is no known study on the influence of sample differences on cone calorimetry data and the degree of the relevant influence. Thus, the subsequent data were reanalyzed using an MCC experiment.

Table 1. CONE test results of CaAlg and CaAlg/nano-Cu_2O. HRR = heat release rate; THR = total heat of release; PHRR = peak heat release rate; EHC = effective heat of combustion; SEA = specific extinction area.

Samples	TTI (s)	HRR (kW/m^2)	THR (MJ/m^2)	PHRR (kW/m^2)	EHC (MJ/kg)	Mean SEA (m^2/kg)
CaAlg	11	39.73	6.40	93.59	10.87	41.46
CaAlg/nano-Cu_2O	6	16.09	2.02	67.30	9.14	21.43

There are two quality-independent parameters in the CONE test—effective heat of combustion (EHC) and mean specific extinction area (mean SEA)—which are reliable results worthy of attention. EHC represents the ratio of HRR to mass loss rate, reflecting the degree of combustion of volatile gases in the gas phase flame. By comparison, it can be found that CaAlg/nano-Cu_2O had a lower EHC value than that of CaAlg, indicating that the overall response of the former was lower than that of the latter, which indirectly proved that the hybrid material had better thermal stability. SEA is defined as the ability of a sample to decompose volatiles produced by flammable masses of combustibles. By comparing the mean SEA of the two materials, it can be clearly seen that CaAlg/nano-Cu_2O exhibited better prominent smoke suppression ability than CaAlg.

Table 2 shows the results of the LOI and UL-94 tests for the two materials. In the UL-94 test, both materials were classified as V-0 based on the fact that both of them exhibited smoldering and maintained their original shape without melt dripping. For the limiting oxygen experiment, the materials are considered to be flame retardant when the LOI value surpasses 27.0. The LOI value of CaAlg was 49.2, which can be already recognized as well flame-retardant. Comparatively, the LOI value of CaAlg/nano-Cu_2O reached 54.0, reflecting a successful improvement by introducing the nano-Cu2O particles to CaAlg.

Table 2. Limiting oxygen index (LOI) test, UL-94 test and MCC test results of CaAlg and CaAlg/nano-Cu_2O.

Samples	LOI (%)	UL-94	HRC (J/g/K)	PHRR (W/g)	THR (kJ/g)	Residues (%)
CaAlg	49.2	V-0	18	16.19	4.3	55.65
CaAlg/nano-Cu_2O	54.0	V-0	16	14.01	4	60.55

MCC, also known as pyrolysis combustion flow calorimetry, is a standard ASTM test method (ASTM 7309) for the analysis of solid materials. It can dynamically measure the HRR and other relevant data based on the oxygen consumed during the experiment [24]. The material can be tested and analyzed at milligram sizes, which can effectively avoid the influence caused by differences in sample preparation [46].

As seen in Table 2, the MCC results are consistent with the phenomenon of the flame experiments. CaAlg had a HRC of 18 J/g/K, a PHRR of 16.19 W/g, and a THR of 4.3 kJ/g, all of which were at lower levels. The fabrication of CaAlg/nano-Cu_2O further improved the flame retardancy of CaAlg, and the values of several parameters declined, especially for the PHRR, which was decreased by nearly 15%. At the same time, the 5% increase in residual carbon confirmed the effect of the nanoparticles on the catalytic carbon formation of CaAlg/nano-Cu_2O as well. It is clear that nano-Cu_2O showed positive influence on the flame retardancy of the hybrid materials.

3.4. Flame-Retardant Mechanism

Scheme 2 shows the flame-retardant mechanism of nano-Cu_2O on the CaAlg/nano-Cu_2O hybrid material in detail. Due to the presence of inorganic nanoparticles, the hybrid materials formed a three-dimensional network structure with low thermal conductivity, which showed superior thermal stability at the initial stage of combustion.

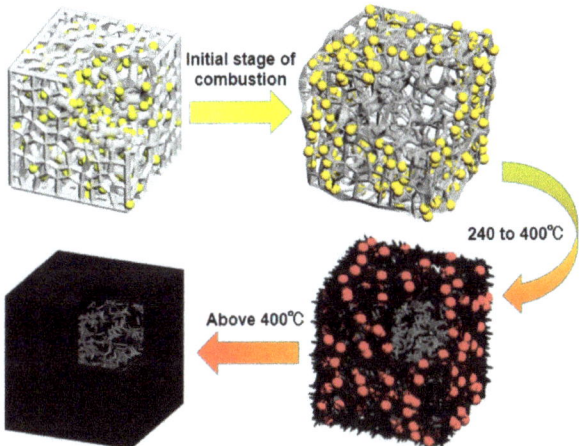

Scheme 2. Flame-retardant mechanism of nano-Cu_2O on the hybrid materials.

At 40–400 °C, the skeleton of CaAlg gradually collapsed and disintegrated with the increase of temperature. The nano-Cu_2O attached to the skeleton had a high surface energy and rapidly migrated and deposited on the surface of CaAlg/nano-Cu_2O. Since Cu_2O catalyzed the thermal decomposition reaction of CaAlg, the outermost CaAlg is the first to be decomposed at a relatively low temperature. Vapor and CO_2 generated by dehydration and decarboxylation in the CaAlg thermal decomposition process carried away most of the heat, while the remaining solid residue, including CuO oxidized from Cu_2O, formed a dense carbon barrier in the outermost layer of CaAlg/nano-Cu_2O. The barrier layer prevented heat and flammable gas from diffusing into the inner layer, while the release of the incombustible gas lowered the concentration of the flammable gas at both the inner and outer contact surfaces, suppressing the overall thermal decomposition reaction to a lower degree and greatly reducing the HRR.

Above 400 °C, the thermal decomposition product $CaCO_3$ began to shift to CaO. As the temperature increased further, char formation began. At this time, the CuO in the calcination residue of CaAlg/nano-Cu_2O became a catalyst for the char formation reaction. A large amount of the carbonaceous carbon produced was coated on the surface of CaAlg/nano-Cu_2O and appeared to be an expansion of the calcination residue. It promoted a denser barrier layer and a narrower surface void that both shielded the diffusion of flammable gases and blocked heat and heat radiation. The heat feedback of the material was reduced, and thereby the progress of the combustion reaction was suppressed. This further improved the flame retardancy of the hybrid materials.

4. Conclusions

In this study, nano-cuprous oxide was prepared by employing Cu^{2+} complexes as precursors via a non-chloride in situ method rather than the widely used direct crosslinking of the sodium alginate solution template. Then, the composites based on the calcium alginate/cuprous oxide nanoparticles were fabricated by a simple green crosslinking and freeze-drying process. Measurements and characterizations show that CaAlg/nano-Cu_2O displayed a completely different three-dimensional network structure from CaAlg, with a LOI value up to 54.0. The addition of a mere 1 wt % nano-Cu_2O reduced the PHHR by 15% and increased the residual carbon rate by 5%, which contributed to the improvement for the flame retardancy of CaAlg/nano-Cu_2O. Moreover, it can be inferred from the supposed working and flame-retardant mechanism of nano-Cu_2O on CaAlg that the carbon layer that formed at low temperature due to catalysis protected the internal material, reduced the HRR, THR, PHRR and EHC, and suppressed the release of smoke of the hybrid materials. The char formation was reinforced by CuO catalysis to reduce combustion performance and improve the flame-retardant properties of the material. CaAlg/nano-Cu_2O not only has potential development prospects in the field of antibacterial and catalytic materials, but also a green material with lower heat release and smoke release, and improved flame-retardant properties compared to CaAlg.

Author Contributions: Conceptualization, Z.L. and Q.L.; Data curation, P.S.; Investigation, P.S., P.X. and L.Z.; Resources, Y.X. and X.Z.; Writing—original draft, P.S.; Writing—review and editing, Z.L.

Funding: This work was supported by the National Natural Science Foundation of China (grant number 51773102).

Acknowledgments: The authors wish to thank Tingting Zhang from Biology Experimental Teaching Center of College of Life Sciences in Qingdao University for the use of some instruments and assistance with the measurements.

Conflicts of Interest: The authors declare no conflict of interest.

References

1. Sun, S.; Zhang, X.; Yang, Q.; Liang, S.; Zhang, X.; Yang, Z. Cuprous oxide (Cu_2O) crystals with tailored architectures: A comprehensive review on synthesis, fundamental properties, functional modifications and applications. *Prog. Mater. Sci.* **2018**, *96*, 111–173. [CrossRef]
2. Qiu, S.L.; Xing, W.Y.; Feng, X.M.; Yu, B.; Mu, X.W.; Yuen, R.K.K.; Hu, Y. Self-standing cuprous oxide nanoparticles on silica@polyphosphazene nanospheres: 3D nanostructure for enhancing the flame retardancy and toxic effluents elimination of epoxy resins via synergistic catalytic effect. *Chem. Eng. J.* **2017**, *309*, 802–814. [CrossRef]
3. Aguilar, M.S.; Rosas, G. A new synthesis of Cu_2O spherical particles for the degradation of methylene blue dye. *Environ. Nanotechnol. Monit. Manag.* **2019**, *11*, 100195. [CrossRef]
4. Su, R.; Li, Q.; Chen, Y.; Gao, B.; Yue, Q.; Zhou, W. One-step synthesis of Cu_2O@carbon nanocapsules composites using sodium alginate as template and characterization of their visible light photocatalytic properties. *J. Clean. Prod.* **2019**, *209*, 20–29. [CrossRef]

5. Kumar, S.; Ojha, A.K.; Bhorolua, D.; Das, J.; Kumar, A.; Hazarika, A. Facile synthesis of CuO nanowires and Cu$_2$O nanospheres grown on rGO surface and exploiting its photocatalytic, antibacterial and supercapacitive properties. *Phys. B* **2019**, *558*, 74–81. [CrossRef]
6. Kociolek-Balawejder, E.; Stanislawska, E.; Dworniczek, E.; Seniuk, A.; Jacukowicz-Sobala, I.; Winiarska, K. Cu$_2$O doped gel-type anion exchanger obtained by reduction of brochantite deposit and its antimicrobial activity. *React. Funct. Polym.* **2019**, *141*, 42–49. [CrossRef]
7. Yang, Z.; Hao, X.; Chen, S.; Ma, Z.; Wang, W.; Wang, C.; Yue, L.; Sun, H.; Shao, Q.; Murugadoss, V.; et al. Long-term antibacterial stable reduced graphene oxide nanocomposites loaded with cuprous oxide nanoparticles. *J. Colloid Interface Sci.* **2019**, *533*, 13–23. [CrossRef]
8. Wang, D.; Kan, Y.; Yu, X.; Liu, J.; Song, L.; Hu, Y. In situ loading ultra-small Cu$_2$O nanoparticles on 2D hierarchical TiO$_2$-graphene oxide dual-nanosheets: Towards reducing fire hazards of unsaturated polyester resin. *J. Hazard. Mater.* **2016**, *320*, 504–512. [CrossRef]
9. Guo, Y.; Dai, M.M.; Zhu, Z.X.; Chen, Y.Q.; He, H.; Qin, T.H. Chitosan modified Cu$_2$O nanoparticles with high catalytic activity for p-nitrophenol reduction. *Appl. Surf. Sci.* **2019**, *480*, 601–610. [CrossRef]
10. Jing, G.J.; Zhang, X.J.; Zhang, A.A.; Li, M.; Zeng, S.H.; Xu, C.J.; Su, H.Q. CeO$_2$-CuO/Cu$_2$O/ Cu monolithic catalysts with three-kind morphologies Cu$_2$O layers for preferential CO oxidation. *Appl. Surf. Sci.* **2018**, *434*, 445–451. [CrossRef]
11. Shang, K.; Liao, W.; Wang, J.; Wang, Y.T.; Wang, Y.Z.; Schiraldi, D.A. Nonflammable alginate nanocomposite aerogels prepared by a simple freeze-drying and post-cross-linking method. *ACS Appl. Mater. Interfaces* **2016**, *8*, 643–650. [CrossRef] [PubMed]
12. Lee, K.Y.; Mooney, D.J. Alginate: Properties and biomedical applications. *Prog. Polym. Sci.* **2012**, *37*, 106–126. [CrossRef] [PubMed]
13. Aime, C.; Coradin, T. Nanocomposites from biopolymer hydrogels: Blueprints for white biotechnology and green materials chemistry. *J. Polym. Sci. Part B Polym. Phys.* **2012**, *50*, 669–680. [CrossRef]
14. Chen, H.B.; Shen, P.; Chen, M.J.; Zhao, H.B.; Schiraldi, D.A. Highly efficient flame retardant polyurethane foam with alginate/clay aerogel coating. *ACS Appl. Mater. Interfaces* **2016**, *8*, 32557–32564. [CrossRef]
15. Safaei, M.; Taran, M.; Imani, M.M. Preparation, structural characterization, thermal properties and antifungal activity of alginate-CuO bionanocomposite. *Mater. Sci. Eng. C* **2019**, *101*, 323–329. [CrossRef]
16. Zhai, Z.; Ren, B.; Xu, Y.; Wang, S.; Zhang, L.; Liu, Z. Green and facile fabrication of Cu-doped carbon aerogels from sodium alginate for supercapacitors. *Org. Electron.* **2019**, *70*, 246–251. [CrossRef]
17. Carneiro-Da-Cunha, M.G.; Cerqueira, M.A.; Souza, B.W.S.; Carvalhoc, S.; Quintas, M.A.C.; Teixeira, J.A.; Vicente, A.A. Physical and thermal properties of a chitosan/alginate nanolayered PET film. *Carbohydr. Polym.* **2010**, *82*, 153–159. [CrossRef]
18. Grant, G.T.; Morris, E.R.; Rees, D.A.; Smith, P.J.C.; Thom, D. Biological interactions between polysaccharides and divalent cations: The egg-box model. *FEBS Lett.* **1973**, *32*, 195–198. [CrossRef]
19. Serrano-Aroca, A.; Ruiz-Pividal, J.F.; Llorens-Gamez, M. Enhancement of water diffusion and compression performance of crosslinked alginate films with a minuscule amount of graphene oxide. *Sci. Rep.* **2017**, *7*, 11684. [CrossRef]
20. Das, D.; Bang, S.; Zhang, S.M.; Noh, I. Bioactive Molecules release and cellular responses of alginate-tricalcium phosphate particles hybrid gel. *Nanomaterials* **2017**, *7*, 389. [CrossRef]
21. Matricardi, P.; Pontoriero, M.; Coviello, T.; Casadei, M.A.; Alhaique, F. In situ cross-linkable novel alginate-dextran methacrylate IPN hydrogels for biomedical applications: Mechanical characterization and drug delivery properties. *Biomacromolecules* **2008**, *9*, 2014–2020. [CrossRef] [PubMed]
22. Siqueira, P.; Siqueira, E.; de Lima, A.E.; Siqueira, G.; Pinzon-Garcia, A.D.; Lopes, A.P.; Segura, M.E.C.; Isaac, A.; Pereira, F.V.; Botaro, V.R. Three-dimensional stable alginate-nanocellulose gels for biomedical applications: Towards tunable mechanical properties and cell growing. *Nanomaterials* **2019**, *9*, 78. [CrossRef] [PubMed]
23. Liu, Y.; Zhao, X.R.; Peng, Y.L.; Wang, D.; Yang, L.W.; Peng, H.; Zhu, P.; Wang, D.Y. Effect of reactive time on flame retardancy and thermal degradation behavior of bio-based zinc alginate film. *Polym. Degrad. Stab.* **2016**, *127*, 20–31. [CrossRef]
24. Liu, Y.; Zhang, C.J.; Zhao, J.C.; Guo, Y.; Zhu, P.; Wang, D.Y. Bio-based barium alginate film: Preparation, flame retardancy and thermal degradation behavior. *Carbohydr. Polym.* **2016**, *139*, 106–144. [CrossRef] [PubMed]

25. Zhang, J.; Ji, Q.; Wang, F.; Tan, L.; Xia, Y. Effects of divalent metal ions on the flame retardancy and pyrolysis products of alginate fibers. *Polym. Degrad. Stab.* **2012**, *97*, 1034–1040. [CrossRef]
26. Liu, Y.; Li, Z.F.; Wang, J.S.; Zhu, P.; Zhao, J.C.; Zhang, C.J.; Guo, Y.; Jin, X. Thermal degradation and pyrolysis behavior of aluminum alginate investigated by TG-FTIR-MS and Py-GC-MS. *Polym. Degrad. Stab.* **2015**, *118*, 59–68. [CrossRef]
27. Ma, X.M.; Li, R.; Zhao, X.H.; Ji, Q.; Xing, Y.C.; Sunarso, J.; Xia, Y.Z. Biopolymer composite fibres composed of calcium alginate reinforced with nanocrystalline cellulose. *Compos. Part A Appl. Sci. Manuf.* **2017**, *96*, 155–163. [CrossRef]
28. Liu, Z.Q.; Li, Z.; Yang, Y.X.; Zhang, Y.L.; Wen, X.; Li, N.; Fu, C.; Jian, R.K.; Li, L.J.; Wang, D.Y. A geometry effect of carbon nanomaterials on flame retardancy and mechanical properties of ethylene-vinyl acetate/magnesium hydroxide composites. *Polymers* **2018**, *10*, 1028. [CrossRef]
29. Wu, N.J.; Niu, F.K.; Lang, W.C.; Xia, M.F. Highly efficient flame-retardant and low-smoke-toxicity poly (vinyl alcohol)/alginate/montmorillonite composite aerogels by two-step crosslinking strategy. *Carbohyd. Polym.* **2019**, *221*, 221–230. [CrossRef]
30. Lewin, M. Some comments on the modes of action of nanocomposites in the flame retardancy of polymers. *Fire Mater.* **2003**, *27*, 1–7. [CrossRef]
31. Li, J.; Li, Z.; Zhao, X.; Deng, Y.; Xue, Y.; Li, Q. Flame retardancy and thermal degradation mechanism of calcium alginate/$CaCO_3$ composites prepared via in situ method. *J. Therm. Anal. Calorim.* **2017**, *131*, 2167–2177. [CrossRef]
32. Liu, Z.; Li, Z.; Zhao, X.; Zhang, L.; Li, Q. Highly efficient flame retardant hybrid composites based on calcium alginate/nano-calcium borate. *Polymers* **2018**, *10*, 625. [CrossRef] [PubMed]
33. Xu, P.; Shao, P.Y.; Zhang, Q.; Cheng, W.; Li, Z.C.; Li, Q. A novel inherently flame-retardant composite based on zinc alginate/nano-Cu_2O. *Polymers* **2019**, *11*, 1575. [CrossRef] [PubMed]
34. Zhao, W.; Qi, Y.; Wang, Y.; Xue, Y.; Xu, P.; Li, Z.; Li, Q. Morphology and thermal properties of calcium alginate/reduced graphere oxide composites. *Polymers* **2018**, *10*, 990. [CrossRef] [PubMed]
35. Liu, Z.; Li, J.; Zhao, X.; Li, Z.; Li, Q. Surface coating for flame retardancy and pyrolysis behavior of polyester fabric based on calcium alginate nanocomposites. *Nanomaterials* **2018**, *8*, 875. [CrossRef]
36. Zhang, C.J.; Liu, Y.H.; Wang, L.; Zhu, P. Properties of Cu^{2+} modified calcium alginate fiber. *Text. Aux.* **2011**, *1*, 4.
37. Zhang, H.Y.; Liu, C.J.; Chen, L.; Dai, B. Control of ice crystal growth and its effect on porous structure of chitosan cryogels. *Chem. Eng. Sci.* **2019**, *201*, 50–57. [CrossRef]
38. Yang, J.R.; Li, Y.Q.; Liu, Y.B.; Li, D.X.; Zhang, L.; Wang, Q.G.; Xiao, Y.M.; Zhang, X.D. Influence of hydrogel network microstructures on mesenchymal stem cell chondrogenesis in vitro and in vivo. *Acta Biomater.* **2019**, *91*, 159–172. [CrossRef]
39. Hou, X.; Xue, Z.; Xia, Y.; Qin, Y.; Zhang, G.; Liu, H.; Li, K. Effect of SiO_2 nanoparticle on the physical and chemical properties of eco-friendly agar/sodium alginate nanocomposite film. *Int. J. Biol. Macromol.* **2019**, *125*, 1289–1298. [CrossRef]
40. Hou, X.B.; Xue, Z.X.; Xia, Y.Z. Preparation of a novel agar/sodium alginate fire-retardancy film. *Mater. Lett.* **2018**, *233*, 274–277. [CrossRef]
41. Kashiwagi, T.; Du, F.M.; Douglas, J.F.; Winey, K.I.; Harris, R.H.; Shields, J.R. Nanoparticle networks reduce the flammability of polymer nanocomposites. *Nat. Mater.* **2005**, *4*, 928–933. [CrossRef] [PubMed]
42. Papageorgiou, S.K.; Kouvelos, E.P.; Favvas, E.P.; Sapalidis, A.A.; Romanos, G.E.; Katsaros, F.K. Metal-carboxylate interactions in metal-alginate complexes studied with FTIR spectroscopy. *Carbohyd. Res.* **2010**, *345*, 469–473. [CrossRef] [PubMed]
43. Ramdani, Y.; Liu, Q.; Huiquan, G.; Liu, P.; Zegaoui, A.; Wang, J. Synthesis and thermal behavior of Cu_2O flower-like, Cu_2O-C_{60} and Al/Cu_2O-C_{60} as catalysts on the thermal decomposition of ammonium perchlorate. *Vacuum* **2018**, *153*, 277–290. [CrossRef]
44. Shi, R.; Tan, L.; Zong, L.; Ji, Q.; Li, X.; Zhang, K.; Cheng, L.; Xia, Y. Influence of Na (+) and Ca (+) on flame retardancy, thermal degradation, and pyrolysis behavior of cellulose fibers. *Carbohyd. Polym.* **2017**, *157*, 1594–1603. [CrossRef]

45. Dong, Q.X.; Liu, M.M.; Ding, Y.F.; Wang, F.; Gao, C.; Liu, P.; Wen, B.; Zhang, S.M.; Yang, M.S. Synergistic effect of DOPO immobilized silica nanoparticles in the intumescent flame retarded polypropylene composites. *Polym. Adv. Technol.* **2013**, *24*, 732–739. [CrossRef]
46. Yang, C.Q.; He, Q.L.; Lyon, R.E.; Hu, Y. Investigation of the flammability of different textile fabrics using micro-scale combustion calorimetry. *Polym. Degrad. Stab.* **2010**, *95*, 108–115. [CrossRef]

© 2019 by the authors. Licensee MDPI, Basel, Switzerland. This article is an open access article distributed under the terms and conditions of the Creative Commons Attribution (CC BY) license (http://creativecommons.org/licenses/by/4.0/).

MDPI
St. Alban-Anlage 66
4052 Basel
Switzerland
Tel. +41 61 683 77 34
Fax +41 61 302 89 18
www.mdpi.com

Polymers Editorial Office
E-mail: polymers@mdpi.com
www.mdpi.com/journal/polymers

www.ingramcontent.com/pod-product-compliance
Lightning Source LLC
LaVergne TN
LVHW070604100526
838202LV00012B/563